Frontiers of Environmental Law Enforcement

Mohamed Mliless ·
Mamdooh Abdelhameed Abdelmottlep ·
Hector R. Garcia
Editors

Frontiers of Environmental Law Enforcement

Innovations, Challenges, and Global Responses

Springer

Editors
Mohamed Mliless
Independent Researcher
Meknes, Morocco

Mamdooh Abdelhameed Abdelmottlep
University of St. Thomas - Texas
Houston, TX, USA

Hector R. Garcia
Southern New Hampshire University
Manchester, NH, USA

ISBN 978-3-032-10204-1 ISBN 978-3-032-10205-8 (eBook)
https://doi.org/10.1007/978-3-032-10205-8

This Springer imprint is published by the registered company Springer Nature Switzerland AG
The registered company address is: Gewerbestrasse 11, 6330 Cham, Switzerland

Foreword

The enforcement of environmental law has moved from the edges of policy debates to the very center of global concern. What was once discussed mainly in courtrooms, ministries, and law enforcement institutions has become a shared responsibility and a vital pillar in the struggle against ecological destruction that threatens the conditions of life itself. From the bleaching of coral reefs to the trafficking of endangered species, from industrial pollution to the weaponization of natural resources, the scale of harm is staggering, challenging both societies and their legal systems to respond.

This volume, *Frontiers of Environmental Law Enforcement: Innovations, Challenges, and Global Responses*, seeks to offer both a wide-angle view and a practical guide to these urgent questions. It opens with a broad reflection on the shifting horizons of enforcement (*Mohamed Mliless, Mamdooh Abdelmottlep, and Hector R. Garcia*), a chapter that traces the evolving terrain of environmental protection in a world of accelerating risks.

On the theoretical dimension, education and prevention emerge as equally vital. A contribution on teaching enforcement (*Mamdooh Abdelmottlep*) illustrates how legal literacy and environmental awareness can be woven into curricula, preparing future generations to view enforcement not as mere punishment but as an act of civic responsibility. Complementing this vision, another chapter on prevention through education and public policy (*Mihaela-Corina Bucur*) underscores how communities themselves can become active guardians of their environments. Both chapters remind us that enforcement is not only reactive but also formative, shaping the ethical imagination of future leaders and embedding environmental responsibility into the very fabric of civic life.

Building on this conceptual ground, other authors show how forensic science and digital technologies, from the reconstruction of environmental crime scenes (*Lakshmi Priya, Bindu Achathuparambil Gopalakrishnan, and George Kuttiparichel Varghese*) to the use of artificial intelligence, are reshaping our ability to trace, document, and prosecute offenses. These methods are not just technical upgrades; they are lifelines of accountability that link acts of destruction directly to those responsible.

The volume does not shy away from the darker side of environmental crime. One chapter (*Hector R. Garcia*) reveals how global criminal networks exploit natural resources, undermining fragile states and destabilizing societies. At the national scale, Morocco's Environmental Police (*Hind Majdoubi*) provide a striking case study of institutional innovation in the Global South, documenting the creation, training, and adaptation of this force in complex governance landscapes. Together, these chapters expose the fragility of institutions yet highlight the possibilities of resilience and renewal.

The international dimension brings further complexity. A chapter on the devastation of war on ecosystems (*Adil Belbal and Abdelali Arroubi*) makes clear that conflict magnifies environmental destruction, often leaving scars deeper than those inflicted on human communities. Another explores the promise and limits of global cooperation in environmental law (*Rabiaa Bourhim*), underscoring the need for solidarity and dialogue in a fragmented world. A further chapter confronts the debate on ecocide (*Sziebig Orsolya Johanna*), challenging readers to consider how law can expand to criminalize the gravest harms to the planet. Together, these contributions illuminate how law must be reimagined if it is to remain a meaningful shield against planetary harm.

Above all, the conclusion that emerges from these chapters is that enforcement is inseparable from justice—justice for ecosystems, for vulnerable communities, and for generations yet to come. The value of this volume lies in its breadth and depth. It bridges science and law, theory and practice, national case studies and international frameworks. I am convinced that the insights gathered in this volume matter because they speak directly to the challenges of tomorrow: how to hold destructive actors accountable, how to train institutions that remain adaptive, and how to embed environmental responsibility into education, governance, and law. It is worth reiterating that future generations stand to gain greatly from this volume. The issues it raises go beyond the transmission of knowledge; they form a framework for survival. For scholars, the book offers a decisive intervention into ongoing academic debates, expanding the scope of environmental law into new theoretical and practical territories. For policymakers, it signals urgent paths for reform, insisting that the failures of today cannot be repeated tomorrow. For the research community, it sets a benchmark for how interdisciplinary, cross-border scholarship can inform and inspire change.

As the chapters make clear, the road ahead demands innovation, solidarity, and courage. Stronger laws and better-equipped agencies are necessary, but so too are partnerships among governments, scientists, educators, and communities. This book is therefore more than a collection of case studies—it is a call to align scholarship, governance, and civic action in the defense of our shared planet. By documenting both challenges and breakthroughs, *Frontiers of Environmental Law Enforcement* stands as a timely and essential contribution to academic debate, an invaluable resource for the research community, and, above all, a testament to the possibilities of courage, solidarity, and justice in protecting the conditions of life itself.

El Houcine Chougrani
Professor of International Law
and International Relations
Faculty of Law
Cadi Ayyad University
Marrakech, Morocco

Acknowledgements

The editors of *Frontiers of Environmental Law Enforcement: Innovations, Challenges, and Global Responses* extend their heartfelt thanks to all those whose efforts made this volume possible. This book is the outcome of dedicated scholarship, collaborative dialogue, and a shared commitment to confronting the legal and ecological challenges of our time.

We are especially grateful to the contributing authors, whose original research and deep expertise have brought together a rich and timely collection of perspectives:

- *Mohamed Mliless*, for setting the stage with a compelling analysis of the shifting terrain of environmental law enforcement in the 21st century.
- *Héctor R. Garcia*, for his critical examination of organized crime's role in transnational environmental offenses.
- *Hind Majdoubi*, for her valuable contribution to the institutional role of Morocco's Environmental Police.
- *Lakshmi Priya, Bindu Achathuparambil Gopalakrishnan, and George Kuttiparichel Varghese*, for their insightful treatment of environmental crime scene investigation and forensic methods.
- *Mamdooh Abdelmottlep*, for his chapter on the pedagogy of environmental crime enforcement and the ethical development of law enforcement officers.
- *Adil Belbal and Abdelaali Arroubi*, for their probing analysis of environmental protection during armed conflict.
- *Rabiaa Bourhim*, for his legal critique of international enforcement challenges and opportunities.
- *SZIEBIG Orsolya Johanna*, for her forward-thinking chapter on the enforceability of ecocide in domestic and international law.
- *Mihaela-Corina Bucur*, for concluding the volume with a much-needed reflection on the role of public policy and environmental education in crime prevention.

We wish to acknowledge, in particular, the collective editorial effort that shaped this volume. The editors, *Mohamed Mliless, Mamdooh Abdelmottlep*, and *Héctor R. Garcia*, worked in close coordination to curate, review, and refine each chapter. Their responsibilities extended far beyond initial selection and organization; they devoted significant time and care to ensuring coherence across contributions, providing in-depth editorial feedback, managing revisions, and overseeing final proofreading and formatting. Their joint effort reflects a shared vision for a volume that is both academically rigorous and practically relevant to contemporary environmental challenges.

We are also indebted to the contributors who peer-reviewed the chapters and offered thoughtful critiques during the development process. Their feedback contributed meaningfully to the clarity and quality of the final texts.

Our thanks go to the editorial and production team at Springer, whose support and professionalism were vital throughout.

This volume is dedicated to the growing community of researchers, advocates, policymakers, and educators committed to building a more just, responsive, and ecologically grounded future. May the insights shared in these pages serve as both reflection and provocation for the work that lies ahead.

The Editors
Mohamed Mliless
Mamdooh Abdelhameed Abdelmottlep
Hector R. Garcia

General Introduction

The volume *Frontiers of Environmental Law Enforcement: Innovations, Challenges, and Global Responses*, edited by Mohamed Mliless and Mamdooh Abdelhameed Abdelmottlep, examines the evolving landscape of environmental enforcement at a time when ecological harm is not incidental but structural, global, and deliberate. Bringing together perspectives from law, criminology, ecology, governance, pedagogy, and forensic science, the book positions enforcement as both a legal necessity and a moral imperative.

The opening chapters set the technical and methodological foundations for enforcement. They show how environmental crime scene investigation, digital forensics, and artificial intelligence are reshaping the capacity to detect, reconstruct, and prosecute environmental offenses with evidentiary precision. By foregrounding science and technology, the volume underscores that effective enforcement today depends on interdisciplinary innovation.

From this conceptual groundwork, the book turns to the global dynamics of transnational organized crime. Contributors map illicit economies—illegal logging, wildlife trafficking, waste dumping—that transcend borders and exploit regulatory weaknesses, with particular attention to their impact on vulnerable regions of the Global South. The discussion then narrows to the national level with a case study of Morocco's Environmental Police, highlighting institutional innovation within a complex governance framework and the challenges of translating legal authority into effective oversight.

Education and pedagogy emerge as vital dimensions of enforcement. One chapter explores how curricula can cultivate awareness, ethics, and civic responsibility, equipping future generations of officers, policymakers, and citizens to integrate enforcement with broader goals of ecological justice.

Extraordinary contexts are also addressed. The chapter on armed conflict exposes how wars devastate ecosystems as profoundly as they damage human societies, and how existing humanitarian law remains inadequate in safeguarding nature. Building outward, the discussion of green international law and the proposed crime of ecocide situates enforcement within global legal reform, emphasizing the need for stronger frameworks to ensure accountability for large-scale environmental destruction.

The final contributions highlight prevention and education as indispensable complements to punitive mechanisms. They stress that enforcement must extend beyond sanctioning violators to fostering cultural and policy shifts that make sustainability and ecological responsibility integral to social life.

Taken together, these contributions demonstrate that environmental enforcement cannot be reduced to the narrow application of statutes. It is a dynamic, contested, and evolving process—one that integrates science with governance, pedagogy with justice, and national institutions with global frameworks. This volume thus advances the argument that enforcement is inseparable from ecological justice and indispensable for building a sustainable future.

Aims and Importance of the Volume

The aim of this volume is to address the widening gap between the scale of environmental harm and the capacity of current enforcement systems. By bringing together insights from multiple disciplines and contexts, it highlights innovations, identifies persistent challenges, and proposes alternative futures where enforcement is more effective, just, and ecologically grounded.

For practitioners, the volume offers comparative insights into institutional models, legal frameworks, and investigative tools that can strengthen enforcement in practice. For teachers, it provides pedagogical models and case studies to enrich curricula in law, criminology, and environmental studies. For students, it presents accessible yet rigorous analyses of emerging debates, preparing them to engage with enforcement as both a legal and ethical endeavor. For researchers, it opens new lines of inquiry by combining doctrinal legal analysis with empirical, forensic, and educational approaches.

Above all, this collection insists that enforcement must go beyond punitive logic to embody an ethic of ecological care and justice. It repositions enforcement not as an isolated legal exercise, but as part of broader struggles for sustainability, human rights, and ecological survival.

Thematic Overview

Nine chapters that examine different but connected facets of environmental law enforcement are collected in this volume. The chapters address the ethical, legal, and political challenges of preserving the environment in the face of rapidly increasing degradation. Even though the contexts and approaches differ—from criminal cartography and institutional design to forensic reconstruction, international legal development, and environmental law enforcement pedagogy—the book's framework aims to lead the reader from general systemic viewpoints to more focused institutional and legal answers, ending with inspiring thoughts on reform, prevention, and education.

In Chapter 1, *Navigating the New Frontiers of Environmental Law Enforcement: The Evolving Landscape of Environmental Protection*, Mohamed Mliless, Mamdooh Abdelmottlep, and Hector R. Garcia set the tone for the entire volume. Their contribution frames environmental law enforcement as a shifting terrain shaped by geopolitical instability, the rapid evolution of environmental crime, and the emergence of new technological tools. This chapter focuses on the evolving architecture of environmental law enforcement in an era defined by intensifying ecological crises and accelerating technological innovation. Against the backdrop of climate disruption, biodiversity loss, and escalating transnational environmental crime, this chapter situates enforcement at the intersection of three constitutive forces: technological advancement, the rising influence of non-state actors, and the imperatives of global cooperation. The triangular model advanced here conceptualizes these forces as mutually reinforcing yet tension-laden domains, shaping both the possibilities and the vulnerabilities of contemporary enforcement regimes. The chapter first highlights the transformative role of emerging technologies—artificial intelligence, blockchain, satellite imagery, and environmental DNA—in enabling anticipatory surveillance and predictive enforcement, while simultaneously exposing legal systems to new ethical, evidentiary, and governance challenges. It then examines the reconceptualization of enforcement within a multi-stakeholder framework, where corporations, NGOs, civil society, and scientific communities complement and contest state authority through mechanisms such as ESG frameworks, citizen-led monitoring, and private certification. At the global scale, the chapter explores how post-pandemic interdependence underscores the urgency of international coordination in combating transboundary environmental crime, while also revealing persistent asymmetries in legal standards, capacities, and political will. By revisiting emblematic failures, including the Deepwater Horizon disaster, Amazon deforestation, and the toxic legacies of e-waste and industrial pollution, the chapter underscores the limits of fragmented enforcement and the lessons for adaptive governance. Looking forward, it identifies emerging trajectories in enforcement practice, including the shift from reactive regulation to proactive prevention, the integration of market-based and collaborative approaches, and the professional retooling of enforcement personnel in hybrid, interdisciplinary roles. Ultimately, the chapter argues for a resilient, technologically literate, and globally networked enforcement architecture—one capable not only of sanctioning environmental crime but of safeguarding ecological futures in conditions of profound planetary change.

The practicalities of environmental crime investigation are the focus of Chapter 2, *Environmental Crime Scene Investigation: Processes and Practices*, co-authored by Lakshmi Priya, Bindu Achathuparambil Gopalakrishnan, and George Kuttiparichel Varghese. The authors argue that environmental violations often unfold over time and space in ways that challenge traditional investigative models. They propose a shift from the notion of a "crime scene" to that of a "crime scenario," and underscore the importance of forensic tools, spatial modeling, and interdisciplinary cooperation in reconstructing environmental harm and attributing responsibility.

Expanding on the global scale of environmental crime, Chapter 3 by Hector R. Garcia, titled *Tracing the Interrelation Between Organized Crime and Environmental*

Degradation: The Criminal Cartography of Environmental Destruction, maps how illicit networks operate across borders to traffic wildlife, timber, minerals, and toxic waste. Garcia draws on recent UNODC and INTERPOL data to demonstrate how organized crime has embedded itself in fragile economies, particularly in the Global South, exploiting both technological loopholes and institutional vulnerabilities. His chapter places these crimes within a broader context of environmental injustice and calls for more sophisticated enforcement strategies, including AI-based surveillance and transnational legal frameworks. For a legal analysis of how transnational environmental crime is addressed within the framework of international law, see Chapter 7 of this volume.

Turning from global networks to national responses, Chapter 4, *Enhancing Environmental Governance: The Role and Impact of Morocco's Environmental Police*, authored by Hind Majdoubi, provides a case study in institutional innovation. Across the globe, environmental governance has increasingly relied on the creation of specialized institutions to address the growing complexity of ecological challenges. Effective monitoring, enforcement, and regulatory oversight are now recognized as essential components of national strategies to protect ecosystems and public health. Within this context, Morocco has taken a significant step by establishing the Environmental Police in 2017 under the framework of the National Charter for the Environment and Sustainable Development. This chapter explores the emergence of this body as a case study of institutional innovation in the Global South. The aim is to assess its statutory foundations, operational structure, and practical effectiveness in strengthening environmental law enforcement. Methodologically, the chapter draws on an analysis of legislative provisions, ministerial decrees, official reports, and interviews with practitioners to trace both achievements and shortcomings. Findings show that the Environmental Police has enhanced environmental control and judicial inquiry capacities, while also facing persistent constraints such as resource limitations, coordination gaps among agencies, and the need for stronger public communication. The implications extend beyond Morocco, offering insights into how specialized enforcement bodies can contribute to regulatory reform and providing comparative lessons for countries seeking to reinforce environmental governance through institutional innovation.

Education and ethical capacity-building are taken up in Chapter 5 by Mamdooh Abdelmottlep, titled *Environmental Crimes Pedagogy*. Abdelmottlep contends that effective enforcement is inseparable from how enforcement officers are trained. He calls for pedagogical models that embed social justice, ethical reasoning, and community engagement into environmental law training. Drawing from criminological theory and global case studies, this chapter offers a blueprint for rethinking how enforcement professionals are prepared to meet the demands of a changing ecological and legal landscape.

A different facet of vulnerability is explored in Chapter 6, *Environmental Protection During Armed Conflicts: Between Legal Codification and Judicial Practice*, by Adil Belbal and Abdelali Arroubi. This chapter examines the status of environmental protection under international humanitarian law and its application in wartime. The authors sift through treaties and tribunal rulings to assess whether international legal promises to protect nature during conflict hold up in reality. Their analysis highlights critical gaps between legal codification and enforcement on the ground. They call for stronger accountability mechanisms that treat environmental harm as a serious consequence of war rather than as a collateral issue.

The legal obstacles to cross-border environmental enforcement are the focus of Chapter 7, *Green International Law Enforcement: Challenges and Opportunities*, authored by Rabiaa Bourhim. International environmental law, also known as "green international law," presents numerous challenges but also offers significant opportunities to protect the global environment. This constantly evolving field addresses issues such as climate change, biodiversity loss, and transboundary pollution. Its objective is to promote sustainable development, preserve natural resources, and prevent environmental degradation. In this study, the role of international environmental law as a branch of international law in the application and enforcement of law on states is examined. This is particularly relevant given the leading role now played by international non-governmental organizations and individuals. The chapter also considers how this influence has become effective in the application of international environmental law, and consequently in international law in general. It aims to analyze the relationship between soft and hard law, and why states might adopt soft law more frequently to regulate environmental issues. For empirical detail on how organized criminal networks operate across regions, particularly in the Global South.

This legal discussion continues in Chapter 8, *Defeating the Unenforceability of Ecocide*, by Sziebig Orsolya Johanna. Building on renewed interest in the concept of ecocide, Johanna interrogates whether ecocide is poised to become a truly enforceable international crime or remains a symbolic placeholder in legal debates. She examines how different branches of law—from international criminal law to EU directives—are approaching the concept, and how strategic legal creativity might overcome the current limitations in prosecuting ecocidal acts. As previously defined in Chapter 7, *green international law* refers to the corpus of international environmental law. In the present chapter, its relevance lies in how ecocide might be incorporated into, or reshape, this existing legal framework.

Finally, Chapter 9, *Prevention of Environmental Crime in the Context of Promoting Public Policies and Environmental Education*, by Mihaela-Corina Bucur, returns the discussion to the broader cultural and systemic level. Dr. Bucur argues that the multidimensional challenges of environmental crime and the evolving frameworks of environmental policy in Europe, situating these within the broader climate crisis and the urgent pursuit of sustainable development. The chapter begins by highlighting the scale and profitability of ecological crimes—illegal logging, waste trafficking, wildlife smuggling, and e-waste dumping—that increasingly rival traditional organized crime. Through data from Europol, UNEP, and Interpol, the chapter shows how ecocide has become one of the fastest-growing transnational threats, with significant consequences for ecosystems, economies, and human health. Importantly, Dr. Bucur then examines the European Union's environmental policies, focusing on legislative instruments, regulatory frameworks, and strategic initiatives such as the European Green Deal, the Biodiversity Strategy 2030, the Circular Economy Action Plan, and the Zero Pollution Action Plan. In this vein, special attention is given to how these policies operationalize key principles—most notably the precautionary principle and the polluter-pays principle—while addressing pressing concerns across domains such as air quality, plastics, biodiversity, soils, and the urban environment. To illustrate this, Dr. Bucur draws on case studies from EU enforcement operations to illustrate how environmental crimes are policed and prosecuted, with emphasis on transnational cooperation through mechanisms such as EMPACT and the SWIPE project. In addition, she foregrounds the role of environmental education in cultivating eco-social competencies, raising awareness of ecological crises, and embedding values of empathy, responsibility, and civic participation into formal and informal learning. By linking education to legislative reform and governance strategies, this chapter underscores the necessity of integrating climate change education into broader social and political responses to ecological crime. Finally, the chapter reviews recent legislative developments in Romania and the EU, with particular emphasis on the expansion of criminal liability to include air pollution under Law no. 56/2025. This marks a significant advance in aligning national and European frameworks while closing long-standing regulatory gaps. Ultimately, the chapter argues that preventing and prosecuting environmental crime requires an integrated approach, combining strong legislation, transnational cooperation, and education-driven cultural transformation to safeguard ecological integrity and human well-being.

Mohamed Mliless
Mamdooh Abdelhameed Abdelmottlep
Hector R. Garcia

25/08/2025

Contents

About the Editors

Dr. Mohamed Mliless is an independent Moroccan scholar specializing in ecolinguistics and critical environmental discourse analysis. He holds a PhD in applied linguistics from Moulay Ismail University in Meknes, Morocco. His research engages deeply with the intersection of language, environment, and social justice, contributing to the emerging field of ecolinguistics with a distinctive focus on Global South perspectives. Dr. Mliless's interdisciplinary work spans ecolinguistics, media literacy, environmental communication, and the linguistic framing of ecological issues. He is the author and coauthor of numerous scholarly publications, including books and peer-reviewed articles that examine environmental discourse, legal rhetoric, and media narratives. Notable among his contributions are chapters in Ecolinguistics, Social Justice, and Sustainability and Environmental Awareness and Governance, which challenge Western-centric paradigms and foreground voices and ecologies from the Global South. His scholarship is recognized for its critical and integrative approach, bridging applied linguistics, environmental studies, and media analysis. Through this work, Dr. Mliless provides valuable insights into the discursive mechanisms shaping environmental awareness, policy, and public perception, particularly in Moroccan and broader MENA contexts. In addition to his research, Dr. Mliless plays an active role in academic publishing. He serves on the editorial boards of several journals, including the Scopus-indexed ARBITRER. He is also a member of the editorial management team and a reviewer for the Scopus-indexed International Journal of

Educational Research Open (Science Direct) and serves as a reviewer for the International Journal of Police Science (IJPS), the International Journal of Research and Innovation in Social Science (IJRISS), and BP International (India).

Dr. Mamdooh Abdelhameed Abdelmottlep is a former law enforcement officer with extensive professional experience in police training, security education, and international academic cooperation. In addition to founding the International Police Science Association (IPSA), Dr. Mamdooh created the World Internal Security and Police Index (WISPI) and Global Police and Security Index (GPSI), which measure the capacity and efficiency of police and security service providers to address internal security issues. Dr. Mamdooh's work has earned him numerous professional honors, including the 2014 CRIMINOLOGIA.IT, awarded by the United Nations, and the 2016 Distinction Medal, awarded by Sharjah's Ruler, His Highness Sheikh Dr. Sultan Al-Qasimi, for the continuous development of Middle Eastern Police Science and International Security Education. His research focuses on police science, law enforcement, and criminal justice administration.

Dr. Hector R. Garcia has served in public safety for over three decades and has held law enforcement positions ranging from patrol, detective, lieutenant, captain, major, deputy chief, and chief. He has lectured extensively on leadership development in the United States, Central America, the Caribbean, Africa, and the Middle East. He has provided professional development sessions for the US Federal Bureau of Investigation National Academy Associates, the US Department of Homeland Security, several international Ministries of Justice and the Interior, and a wide range of other public safety and professional organizations. Dr. Garcia has contributed significantly to international public safety initiatives, working with partners across the globe and the US Department of State. He has extensive experience in higher education and currently serves as a dean of graduate criminal justice studies at one of the largest accredited higher education institutions in the

United States. Dr. Garcia is a dedicated educator, prolific trainer, and founder of the Family Charity Foundation, embodying a philosophy of hard work, lifelong learning, and community service.

Abbreviations

AI	Artificial Intelligence
Art.	Article
ASEAN-WEN	ASEAN Wildlife Enforcement Network
CITES	Convention on International Trade in Endangered Species of Wild Fauna and Flora
CoE	Council of Europe
CPCB	Central Pollution Control Board
CSI	Crime Scene Investigation
CSR	Crime Scenario Reconstruction
CWF	Canadian Wildlife Federation
DETER	Real-Time Deforestation Detection System
DNA	Deoxyribonucleic Acid
DRC	Democratic Republic of Congo
EAPs	Environmental Action Programs
ECHR	European Convention on Human Rights
E-CSI	Environmental Crime Scene Investigation
E-CSR	Environmental Crime Scenario Reconstruction
ECtHR	European Court of Human Rights
EIA	Environmental Investigation Agency
E-MAGIN	E-Waste Management in Ghana through Integrated Governance
EMPACT	European Multidisciplinary Platform Against Criminal Threats
ENMOD	Convention on the Prohibition of Military or Any Other Hostile Use of Environmental Modification Techniques
EPA	Environmental Protection Agency (U.S.)
EPR	Extended Producer Responsibility
ESG	Environmental, Social, and Governance
EU	European Union
EU EEN	European Union Environmental Enforcement Network
EU GDPR	European Union General Data Protection Regulation
EUROPOL	European Union Agency for Law Enforcement Cooperation
FBINAA	FBI National Academy Associates

FLEGT	Forest Law Enforcement, Governance and Trade
GC-MS	Gas Chromatography–Mass Spectrometry
GIS	Geographic Information Systems
GI-TOC	Global Initiative Against Transnational Organized Crime
HPLC	High-Performance Liquid Chromatography
IACtHR	Inter-American Court of Human Rights
IBAMA	Brazilian Institute of Environment and Renewable Natural Resources
ICC	International Criminal Court
ICP-MS	Inductively Coupled Plasma Mass Spectrometry
ICRC	International Committee of the Red Cross
ICTY	International Criminal Tribunal for the former Yugoslavia
ILC	International Law Commission
INEF	International Network of Environmental Forensics
INTERPOL	International Criminal Police Organization
IWWG	International Waste Working Group
KWS	Kenya Wildlife Service
MEE	Ministry of Ecology and Environment (China)
MoEFCC	Ministry of Environment, Forest and Climate Change
MOESP	Ministry of Environment and Spatial Planning (Slovenia)
MRSC	Member of the Royal Society of Chemistry
MSs	Member States
NGO	Non-Governmental Organization
NIH	National Institutes of Health
NIJ	National Institute of Justice
NMT	Nuremberg Military Tribunals
NPR	National Public Radio
OJJDP	Office of Juvenile Justice and Delinquency Prevention
PNP	Philippine National Police
SCMS	School of Communication and Management Studies
SJD	Scientiae Juridicae Doctor (Doctor of Juridical Science)
SNHU	Southern New Hampshire University
STOP	(Not an acronym; used in "STOP Ecocide Foundation")
SWIPE	Successfully Weaponizing Illicit Wildlife Cases in Europe
TCEQ	Texas Commission on Environmental Quality
TOC	Transnational Organized Crime
UN	United Nations
UNEP	United Nations Environment Programme
UNFAO	United Nations Food and Agriculture Organization
UNGA	United Nations General Assembly
UNODC	United Nations Office on Drugs and Crime
USA	United States of America
WWF	World Wildlife Fund

Chapter 1
Navigating the New Frontiers of Environmental Law Enforcement: The Evolving Landscape of Environmental Protection

Mohamed Mliless, Mamdooh Abdelhameed Abdelmottlep, and Hector R. Garcia

1 Introduction

As we stand at the precipice of unprecedented environmental challenges, the need for effective law enforcement mechanisms has never been more critical. The early decades of the 21st century have witnessed an alarming acceleration of environmental degradation, marked by climate change's intensifying impacts and biodiversity loss at scales previously unimagined. This transformation of our natural world has been accompanied by an equally dramatic evolution in the methods and mechanisms of environmental crime, necessitating a fundamental reimagining of how we approach environmental law enforcement.

This volume, "Frontiers of Environmental Law Enforcement: Innovations, Challenges, and Global Responses," emerges at a crucial juncture in human history. It presents a comprehensive examination of how legal systems worldwide are adapting to meet unprecedented environmental challenges while leveraging technological

M. Mliless (✉)
Applied Linguistics, Specializing in Ecolinguistics and Environmental Discourse in the Global South, Meknes, Morocco
e-mail: m.mliless@edu.umi.ac.ma

M. A. Abdelmottlep
Police Science and Law Enforcement, Foreign Legal Consultant, State Bae of Texas, Katy, USA

International Police Science Association (IPSA), Katy, TX, USA

H. R. Garcia
Southern New Hampshire University Global Campus, Criminal Justice Programs, Manchester, USA
e-mail: hrgarcia@pm.me

M. Mliless et al. (eds.), *Frontiers of Environmental Law Enforcement*,
https://doi.org/10.1007/978-3-032-10205-8_1

innovations and responding to emerging global threats. The convergence of technological advancement, shifting geopolitical dynamics, and increasing environmental awareness has created both opportunities and challenges for environmental law enforcement that demand scholarly attention and practical solutions.

2 The Triple Challenge: Technology, Non-state Actors, and Global Cooperation

The contemporary landscape of environmental law enforcement faces three interconnected challenges that define its effectiveness and scope (Hedemann-Robinson, 2019). First, rapid technological advancement presents both opportunities and hurdles in monitoring and enforcement capabilities (White, 2014). Second, the rising influence of non-state actors, from multinational corporations to environmental NGOs, has transformed traditional enforcement paradigms (White, 2014). Third, the inherently global nature of environmental crimes necessitates unprecedented levels of international cooperation (Hedemann-Robinson, 2019). These challenges do not exist in isolation but form a complex web of interactions that shape modern environmental protection efforts. A visual representation of these dynamics can help clarify the deeply interconnected nature of modern environmental law enforcement. The conceptual model in Figure 1.1 illustrates how three critical forces, technological innovation, international cooperation, and transnational criminal activity, interact to influence the success or failure of enforcement efforts. Each corner of the triangle represents a domain that either strengthens or weakens the others, depending on how they are leveraged or neglected. This triangular model reflects the multidirectional pressures enforcement agencies face, underscoring the need for integrated, globally informed responses.

2.1 Technological Innovation in Environmental Law Enforcement

The rapid pace of technological advancement has transformed both the perpetration and prevention of environmental crimes. Artificial Intelligence (AI) (Collins, 2023) and blockchain technology, in particular, have emerged as powerful tools in the environmental law enforcement arsenal (Al-Qassim et al., 2022). In terms of AI, Collins (2023) assumes that innovative AI's capabilities in pattern recognition and predictive analytics are revolutionizing how we detect and prevent environmental violations, from illegal logging to toxic waste dumping. Satellite imagery analysis, powered by machine learning algorithms, now enables real-time monitoring of vast geographical areas, providing enforcement agencies with unprecedented surveillance capabilities (Handan-Nader et al., 2020).

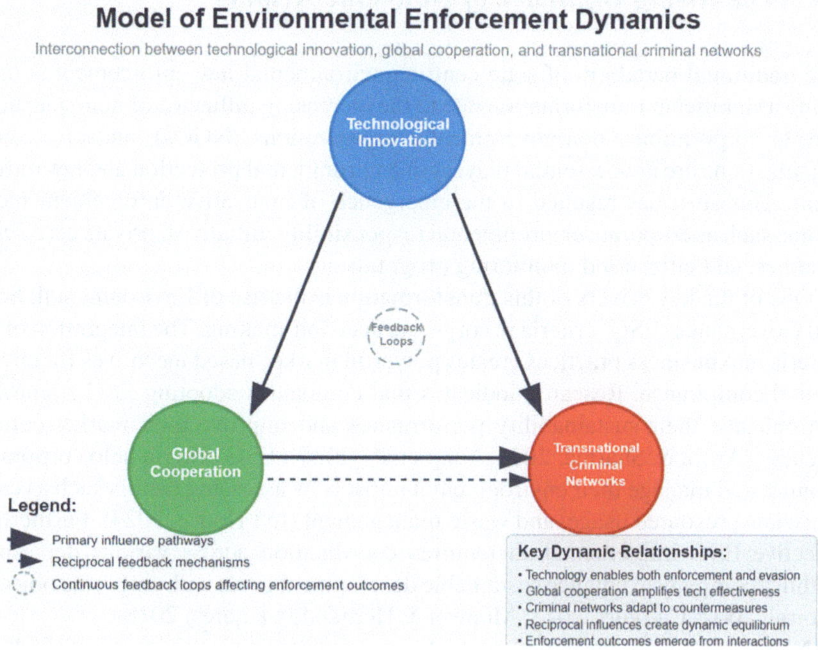

Fig. 1 Triangular model of interconnected environmental enforcement factors

The technological revolution in environmental law enforcement brings its own set of challenges that must be addressed carefully. Questions of data privacy, algorithmic bias, and the digital divide between developed and developing nations must be considered. Ensuring responsible handling of sensitive environmental data and compliance with privacy regulations like the EU GDPR is paramount (Collins, 2023). Algorithmic bias in AI systems can lead to unfair treatment if training data contains societal biases, making it essential to ensure fairness and prevent environmental injustices (Handan-Nader et al., 2020). The digital divide impacts the effectiveness of enforcement, and bridging this gap is crucial for equitable enforcement (Collins, 2023). Integrating these technologies into legal frameworks raises complex questions about the admissibility of digital evidence in court, requiring it to meet legal standards for effective enforcement (Handan-Nader et al., 2020). The global nature of environmental issues often involves multiple jurisdictions, necessitating clear guidelines and cooperation between authorities. While technology enhances efficiency and accuracy, maintaining a balance with human judgment is essential to ensure fair and just outcomes. Addressing these challenges requires collaboration between technologists, legal experts, and policymakers to ensure technological advancements contribute positively to environmental law enforcement.

2.2 The Rising Influence of Non-state Actors

The traditional paradigm of state-centric environmental law enforcement is undergoing a significant transformation due to the increasing influence of non-state actors. Private corporations, non-governmental organizations (NGOs), and civil society organizations are now essential players in environmental protection and law enforcement. This shift has resulted in the emergence of innovative enforcement mechanisms, such as corporate environmental responsibility initiatives, private certification schemes, and citizen-led monitoring programs.

One of the key drivers of this transformation is the rise of Environmental, Social, and Governance (ESG) criteria in corporate decision-making. The integration of ESG criteria into business practices creates powerful market-based incentives for environmental compliance. Research indicates that companies adopting ESG frameworks can enhance their sustainability performance and improve their market competitiveness (Arora & Sharma, 2022; Yang et al., 2024). ESG criteria help corporations monitor and manage their environmental impacts by assessing factors such as carbon emissions, resource usage, and waste management (Lyt House, 2024). Furthermore, effective ESG implementation requires coordination across various departments within a company to balance sustainable development goals with corporate objectives (Boiral, 2002; Çalıyurt, 2022; Moneva & Hernández-Pajares, 2018).

NGOs have also developed sophisticated monitoring and reporting capabilities that often rival those of state agencies. This evolution in monitoring practices has enabled NGOs to hold corporations accountable for their environmental impacts and promote transparency in corporate governance (Cek & Eyupoglu, 2020). The ability of NGOs to engage in citizen-led monitoring programs exemplifies how non-state actors are reshaping the landscape of environmental law enforcement (Haque, 2020).

The reconceptualization of environmental law enforcement within a multi-stakeholder framework is crucial as it reflects a broader understanding of governance that includes diverse perspectives from various actors. This shift acknowledges that effective environmental governance cannot rely solely on state mechanisms but must incorporate the roles played by private entities and civil society (Nonet et al., 2022). As ESG considerations become increasingly integral to corporate strategies, companies that fail to adapt may face reputational risks and diminished access to capital (Simpson Grierson, 2022; Yang et al., 2024).

Beyond NGOs and corporate entities, an expanding range of non-enforcement actors now play indispensable roles in environmental protection. Ecologists provide the scientific assessments and ecological baselines that underpin prosecution efforts. Community leaders often act as local enforcers, sounding the alarm in regions where government presence is minimal or contested. Technologists and software engineers are increasingly integral to the development of tools such as sensor networks, AI-powered monitoring platforms, and environmental data visualization systems. These diverse contributors bring critical expertise and legitimacy to enforcement efforts. Their collaboration with state authorities not only enhances the accuracy of environmental assessments but also builds community trust and fosters multi-level environmental governance.

2.3 *Global Cooperation in a Post-pandemic World*

The COVID-19 pandemic has starkly illustrated the interconnected nature of global challenges and the critical importance of international cooperation (Jörgens et al., 2023). Environmental crimes frequently transcend national boundaries, requiring coordinated responses from multiple jurisdictions (Spapens et al., 2016). The pandemic has also highlighted the links between environmental degradation and public health crises, emphasizing the need for integrated approaches to environmental protection (Hammad et al., 2023).

International cooperation in environmental law enforcement faces numerous challenges, including varying regulatory standards, jurisdictional conflicts, and resource disparities between nations (Norouzi & Ataei, 2021). The pandemic has both exacerbated these challenges and provided valuable lessons in global coordination and rapid response mechanisms (Haysom & Shaw, 2022).

3 Learning from Failures in Environmental Law Enforcement

The history of environmental law enforcement is marked by both catastrophic failures and transformative lessons that continue to shape contemporary practices. While the Deepwater Horizon disaster and Amazon deforestation offer vital lessons, a more global set of examples reveals the widespread nature of enforcement failures. In Ghana's Agbogbloshie region, illegal e-waste dumping—often disguised as donations—has created one of the world's most toxic environments, highlighting loopholes in international waste regulation and poor domestic enforcement capacity. Similarly, the Citarum River in Indonesia, often labeled the most polluted river in the world, reflects how weak industrial regulation and under-resourced environmental agencies can lead to systemic contamination. These cases underscore the reality that environmental crime is not confined to one region or ecosystem, and that enforcement strategies must be shaped by diverse geographic, economic, and institutional contexts.

Through examining high-profile cases such as the Deepwater Horizon disaster (Kotsakis & Boukli, 2023) and ongoing Amazon deforestation (Handan-Nader et al., 2020), we gain critical insights into the complexities and challenges of environmental protection. These cases serve as powerful reminders of the consequences of inadequate enforcement mechanisms and regulatory oversight. Moreover, they highlight the urgent need for robust, adaptive approaches that can effectively address both immediate environmental threats and long-term ecological challenges. By analyzing these failures, we can better understand the pathways toward more effective environmental law enforcement.

3.1 The Deepwater Horizon Disaster: Regulatory Failures and Reform

The 2010 Deepwater Horizon oil spill serves as a stark reminder of the consequences of inadequate environmental law enforcement. This catastrophic event resulted in the largest marine oil spill in history, causing unprecedented environmental damage and economic losses. Analysis of this disaster reveals critical gaps in regulatory oversight, emergency response capabilities, and corporate accountability mechanisms (Allain, 2024; U.S. Environmental Protection Agency, n.d.). The aftermath of Deepwater Horizon led to significant reforms in offshore drilling regulations and enforcement practices. However, questions remain about the adequacy of these reforms and their ability to prevent similar disasters in the future (U.S. Environmental Protection Agency, 2013). This case study provides valuable insights into the challenges of regulating complex industrial operations and the importance of robust enforcement mechanisms.

3.2 Amazon Deforestation: The Challenge of Transnational Environmental Crime

The ongoing challenge of Amazon deforestation exemplifies the complexities of enforcing environmental laws across national boundaries. Despite numerous international agreements and national regulations, illegal deforestation continues at alarming rates (Amazon Conservation Association, n.d.). This case highlights the limitations of traditional enforcement approaches and the need for innovative solutions that address the economic drivers of environmental crime.

The role of technology in monitoring deforestation, coupled with international pressure and market-based initiatives, demonstrates the potential for integrated approaches to environmental law enforcement (UNFAO, 2018). However, it also reveals the persistent challenges of balancing economic development with environmental protection in developing nations (Amazon Conservation Association, n.d.).

4 Emerging Trends and Future Directions

The future of environmental law enforcement stands at a pivotal crossroads, shaped by rapidly evolving technologies, shifting global priorities, and emerging environmental challenges. New approaches are emerging that combine AI, citizen science, and innovative regulatory frameworks to address increasingly complex environmental threats (Bichler & Malm, 2008; Strasser et al., 2019; Ceccaroni et al., 2019; Fritz et al., 2019; Fraisl et al., 2020; Serrao et al., 2020). The application of machine learning algorithms has shown particular promise in detecting environmental violations and predicting high-risk areas for illegal activities (Reichstein et al., 2019). As we look ahead, the integration of blockchain technology, environmental DNA analysis, and advanced satellite monitoring promises to revolutionize how we detect and prevent environmental crimes (Fu et al., 2022; Hajjar et al., 2021; Liu et al., 2007). Environmental DNA monitoring has already demonstrated significant potential in tracking biodiversity loss and illegal wildlife trafficking (Ruppert et al., 2019). These developments, coupled with growing international collaboration and adaptive governance models, signal a transformation in environmental law enforcement strategies (Kanashiro Uehara et al., 2023). Understanding these trends is crucial for developing resilient and effective enforcement mechanisms for the future, particularly as environmental challenges become increasingly interconnected and complex (Newig et al., 2017).

4.1 The Promise of Technological Innovation

The rapid advancement of technology continues to create new opportunities for environmental law enforcement. Emerging technologies such as environmental DNA analysis, autonomous drones, and quantum computing hold promise for enhancing detection and enforcement capabilities. However, the effective integration of these technologies requires careful consideration of legal, ethical, and practical implications (Benami et al., 2020; Glicksman et al., 2020).

Among the most promising developments is the deployment of AI to support anticipatory enforcement strategies. AI-based tools can analyze patterns in environmental data to predict hotspots for illegal activity, such as unregulated logging, overfishing, or poaching. For example, machine learning models are now being trained to detect early signs of deforestation from satellite imagery, enabling intervention before irreversible damage occurs. These predictive systems are not only improving the timeliness of enforcement actions but are also shifting the paradigm from reactive punishment to proactive prevention—an essential transition in the context of accelerating ecological degradation.

4.2 The Evolution of Enforcement Mechanisms

Traditional command-and-control regulatory approaches are being supplemented by more flexible and adaptive enforcement mechanisms. Market-based instruments, voluntary compliance programs, and collaborative enforcement initiatives are becoming increasingly important tools in the environmental law enforcement toolkit (Gunningham & Sinclair, 1999; Short & Toffel, 2010). This evolution reflects a growing recognition of the need for diverse approaches to address complex environmental challenges (Coglianese & Nash, 2014).

As these enforcement models evolve, so too must the roles and capacities of enforcement personnel. Today's environmental officers must be equipped not only with legal knowledge but also with interdisciplinary skills in data interpretation, technological literacy, and ecological assessment. Customs agents, environmental rangers, forensic analysts, and judicial investigators are increasingly working in hybrid teams that integrate scientific evidence with legal action. This shift necessitates ongoing professional training and greater investment in enforcement infrastructure, particularly in developing regions. Without strengthening the human dimension of enforcement, even the most advanced legal tools and technologies will fall short of their potential.

4.3 Building Resilient Enforcement Systems

The experience of recent global crises has highlighted the importance of building resilient enforcement systems capable of adapting to emerging challenges. This includes developing robust international cooperation frameworks, enhancing capacity-building initiatives, and strengthening information-sharing mechanisms among enforcement agencies worldwide (Castaño-Rosa et al., 2022; McFarlane, 2005; Rifki, 2024).

5 Conclusion

The complexity of modern environmental challenges requires an integrated approach to law enforcement—one that weaves together technological innovation, multi-stakeholder engagement, and global cooperation. As this chapter has demonstrated, effective enforcement today is no longer the sole domain of government institutions. It is a collective endeavor that involves front-line personnel, legal professionals, ecologists, community leaders, and technologists alike. The growing use of AI, real-time surveillance, and predictive analytics is reshaping enforcement strategies, while partnerships between states and non-state actors are generating new models of legitimacy and accountability. Together, these developments point to a future in which enforcement is not only more effective but also more equitable and ecologically responsive.

The future of environmental law enforcement lies in our ability to harness technological innovations, engage diverse stakeholders, and foster international cooperation effectively. As we face increasingly complex environmental challenges, the importance of robust and adaptive enforcement mechanisms cannot be overstated. This volume represents a significant step toward understanding and addressing these challenges, providing valuable insights for scholars, practitioners, and policymakers working to protect our shared environmental heritage.

References

Allain, E. (2024, March 29). *Environmental justice and the deepwater horizon oil spill*. Retrieved from https://storymaps.arcgis.com/stories/a104a6672af54aeba74179de2b346cac

Al-Qassim, M., Chen, W., Al-Sartawi, A. (2022). Blockchain in environmental compliance and enforcement. In A. M. A. Musleh Al-Sartawi (Ed.), *Artificial intelligence for sustainable finance and sustainable technology. ICGER 2021. Lecture notes in networks and systems* (p. 423). Springer. https://doi.org/10.1007/978-3-030-93464-4_39

Amazon Conservation Association. (n.d.). *Deforestation in the Amazon*. Retrieved from https://www.amazonconservation.org/the-challenge/threats/

Arora, A., & Sharma, D. (2022). Do environmental, social and governance (ESG) performance scores reduce the cost of debt? Evidence from Indian firms. *Australasian Accounting, Business and Finance Journal, 16*, 4–18. https://doi.org/10.14453/aabfj.v16i5.02

Benami, E., Ho, D.O., & McDonough, A. (2020). *Innovations for environmental compliance: emerging evidence and opportunities*. Retrieved from https://siepr.stanford.edu/publications/policy-brief/innovations-environmental-compliance-emerging-evidence-and-opportunities

Bichler, G., & Malm, A. E. (2008). A social network analysis of the evolution of the environmental criminology and crime analysis (ECCA) symposiums. *Journal of Environmental Studies and Sciences, 38*(1), 1–16.

Boiral, O. (2002). Tacit knowledge and environmental management. *Long Range Planning, 35*, 291–317. https://doi.org/10.1016/S0024-6301(02)00047-X

Çalıyurt, K. T. (2022). *New approaches to CSR*. Springer. https://doi.org/10.1007/978-981-16-9364-9

Castaño-Rosa, R., Pelsmakers, S., Järventausta, H., Poutanen, J., Tähtinen, L., Rashidfarokhi, A., & Toivonen, S. (2022). Resilience in the built environment: Key characteristics for solutions to multiple crises. *Sustainable Cities and Society, 87*, Article 104259. https://doi.org/10.1016/j.scs.2022.104259

Ceccaroni, L., Bibby, J., Roger, E., Flemons, P., Michael, K., Fagan, L., & Oliver, J. L. (2019). Opportunities and risks for citizen science in the age of artificial intelligence. *Citizen Science: Theory and Practice, 4*(1), 1–14. https://doi.org/10.5334/cstp.241

Cek, K., & Eyupoglu, S. (2020). Does environmental, social and governance performance influence economic performance? *Journal of Business Economics and Management, 21*(4), 1165–1184

Coglianese, C., & Nash, J. (2014). *Leveraging the private sector: Management-based strategies for improving environmental performance*. Resources for the Future Press. Retrieved from Management-Based Strategies for Improving Private Sector Environmental Performance

Collins, P. (2023). *The role of artificial intelligence in environmental regulation. LSE Blogs*. Retrieved from https://blogs.lse.ac.uk/politicsandpolicy/the-role-of-artificial-intelligence-in-environmental-regulation/

Fritz, S., See, L., Carlson, T., Haklay, M., Oliver, J. L., Fraisl, D., Mondardini, R., Brocklehurst, M., Abernethy, K., Schade, S., & Wehn, U. (2019). Citizen science and the United Nations sustainable development goals. *Nature Sustainability, 2*, 922–930. https://doi.org/10.1038/s41893-019-0390-3

Fraisl, D., Campbell, J., See, L., Wehn, U., Wardlaw, J., Gold, M., Moorthy, I., Arias, R., & Fritz, S. (2020). Mapping citizen science contributions to the UN sustainable development goals. *Sustainability Science, 15*, 1735–1751. https://doi.org/10.1007/s11625-020-00833-7

Fu, B., Zhang, L., Xu, Z., & Liu, Y. (2022). Coupling human and natural systems for sustainability: Experience from China's Loess Plateau. *Ecological Indicators, 133*, Article 108372. https://doi.org/10.1016/j.ecolind.2021.108372

Glicksman, R. L., Markell, D. L., & Monteleoni, C. (2020). Technological innovation, data analytics, and environmental enforcement. *Ecology Law Quarterly, 44*, 41–88. Retrieved from: https://www.ecologylawquarterly.org/wp-content/uploads/2020/03/Technological-Innovation-Data-Analytics-and-Environmental-Enforcement.pdf

Gunningham, N., & Sinclair, D. (1999). Regulatory pluralism: Designing policy mixes for environmental protection. *Law & Policy, 21*(1), 49–76. Retrieved from https://onlinepubs.trb.org/onlinepubs/PBRLit/GunninghamSinclair.pdf

Hajjar, R., Oldekop, J. A., Cronkleton, P., Newton, P., Russell, A. J. M., & Zhou, W. (2021). A global analysis of the social and environmental outcomes of community forests. *Nature Sustainability, 4*(3), 216–224.

Hammad, H. M., Nauman, H. M. F., & Abbas, F. (2023). Impacts of COVID-19 pandemic on environment, society, and food security. *Environmental Science and Pollution Research, 30*, 99261–99272. https://doi.org/10.1007/s11356-023-25714-1

Handan-Nader, C., Ho, D. E., & Liu, L. Y. (2020). Deep learning with satellite imagery to enhance environmental enforcement. In*Stanford digital repository*. Retrieved from https://purl.stanford.edu/bh005pt4088

Haque, M. S. (2020). A critique of the role of NGOs as partners in governance. *Asia Pacific Journal of Public Administration, 42*(1), 17–32. https://doi.org/10.1080/23276665.2020.1748960

Haysom, S., & Shaw, M. (2022). *An analytic review of past responses to ENVIRONMENTAL CRIME and programming recommendations*. Retreived from https://globalinitiative.net/wp-content/uploads/2022/09/GITOC-EU-An-analytic-review-of-past-responses-to-environmental-crime-and-programming-recommendations.pdf

Hedemann-Robinson, M. (2019). *Enforcement of international environmental law challenges and responses at the international level*. Routledge.

Jörgens, H., Knill, C., & Steinebach Y. (Eds.). (2023). *Routledge handbook of environmental policy*. Routledge

Kanashiro Uehara, T., Chazarin, F., Nakagawa, L., Favareto, A., Tobias, T., Favareto, A., Minani, R., Nugroho, B., Setiajiati, F., Iriyani, S., Buchori, D., & Chiavaroli, C. (2023). 'Glocalizing' land-use and forest governance in the tropics: examining research partnerships and international forest policies affecting Brazil, DRC and Indonesia. *F1000Research, 12*, 125. https://doi.org/10.12688/f1000research.130219.1

Kotsakis, A., & Boukli, A. (2023). Transversal harm, regulation, and the tolerance of oil disasters. *Transnational Environmental Law, 12*(1), 71–94. https://doi.org/10.1017/S2047102522000346

Liu, J., Dietz, T., Carpenter, S. R., Folke, C., Alberti, M., Redman, C. L., Schneider, S. H., Ostrom, E., Pell, A. N., Lubchenco, J., Taylor, W. W., Ouyang, Z., & Deadman, P. (2007). Coupled human and natural systems. *Ambio, 36*(8), 639–649.

Lyt House (2024, May 9). *Understanding ESG criteria: Essential insights for businesses*. Retreived from https://www.lythouse.com/blog/understanding-esg-criteria-essential-insights-for-businesses

McFarlane, J. (2005). Regional and international cooperation in tackling transnational crime, terrorism and the problems of disrupted states. *Journal of Financial Crime, 12*(4), 301–309. https://doi.org/10.1108/13590790510624774

Moneva, J. M., & Hernández-Pajares, J. (2018). Corporate social responsibility performance and sustainability reporting in SMEs: An analysis of owner-managers' perceptions. *International journal of Sustainable Economy, 10*(4), 405–420.

Newig, J., Challies, E., Jager, N. W., Kochskaemper, E., & Adzersen, A. (2017). The environmental performance of participatory and collaborative governance: A framework of causal mechanisms. *Policy Studies Journal, 47*(2), 269–297.

Nonet, G. A. H., Gössling, T., Van Tulder, R. et al. (2022). Multi-stakeholder engagement for the sustainable development goals: Introduction to the special issue. *Journal of Business Ethics, 180*, 945–957. https://doi.org/10.1007/s10551-022-05192-0

Norouzi, N., & Ataei, E. (2021). Covid-19 crisis and environmental law: Opportunities and challenges. *Hasanudd in Law Review, 7*(1), 2442–9880. E-ISSN: 2442-9899

Reichstein, M., Camps-Valls, G., Stevens, B., Jung, M., Denzler, J., & Carvalhais, N. (2019). Deep learning and process understanding for data-driven Earth system science. *Nature, 566*(7743), 195–204.

Rifki, M. (2024). Dynamics of transnational crime law enforcement in the era of globalization challenges, opportunities and prospects towards a safer future. *Proceedings: International Forum Research on Education, Social Sciences Technology and Humanities, 1*(2), 263–268. Retrieved from https://journal.berpusi.co.id/index.php/POE/article/view/878

Ruppert, K. M., Kline, R. J., & Rahman, M. S. (2019). Past, present, and future perspectives of environmental DNA (eDNA) metabarcoding: A systematic review in methods, monitoring, and applications of global eDNA. *Global Ecology and Conservation, 17*, Article e00547.

Serrao, N. R., Doan, T. L., & Bush, A. J. (2020). Combining artificial intelligence and citizen science to improve ecological monitoring. *Patterns, 1*(7), 100143. https://doi.org/10.1016/j.patter.2020.100143

Short, J. L., & Toffel, M. W. (2010). Making self-regulation more than merely symbolic: The critical role of the legal environment. *Administrative Science Quarterly, 55*(3), 361–396. https://doi.org/10.2189/asqu.2010.55.3.361

Simpson Grierson. (2022). *Embracing ESG in corporate decision-making: Trends and implications for directors.* Retrieved from Simpson Grierson Insights.

Spapens, T., White, R., & Huisman, W. (Eds.). (2016). *Environmental crime in transnational context global issues in green enforcement and criminology.* Routledge.

Strasser, B. J., Baudry, J., Mahr, D., Sanchez, G., & Tancoigne, E. (2019). Citizen Science"? Rethinking science and public participation. *Science & Technology Studies, 32*(2), 52–76.

UNFAO. (2018, March 5). Innovative technology: A game-changer for monitoring deforestation and land-use change. Retrieved from https://www.fao.org/redd/news/detail/en/c/1105760/

U.S. Environmental Protection Agency. (n.d.). *Deepwater Horizon – BP Gulf of Mexico Oil Spill.* Retrieved from https://www.epa.gov/enforcement/deepwater-horizon-bp-gulf-of-mexico-oil-spill

U.S. Environmental Protection Agency. (2013, January 3). *Transocean settlement.* Retrieved from https://www.epa.gov/enforcement/transocean-settlement

White, R. (2014). Environmental regulation and law enforcement. In G. Bruinsma & D. Weisburd (Eds.), *Encyclopedia of criminology and criminal justice.* Springer. https://doi.org/10.1007/978-1-4614-5690-2_284

Yang, K., Zhang, T., & Ye, C. (2024). The sustainability of corporate ESG performance: An empirical study. *Sustainability, 16*(6), 2377. https://doi.org/10.3390/su16062377

Dr. Mohamed Mliless is an independent researcher and a leading figure in Ecolinguistics and Critical Environmental Discourse Analysis. He holds a Ph.D. in Applied Linguistics from Moulay Ismail University, Meknes. His work includes influential books and articles on legal and environmental discourse, addressing urgent socio-environmental challenges. Dr. Mliless serves on the editorial boards of several journals, including the Scopus-indexed *ARBITRER*. He is also the editorial manager and reviewer for the Scopus-indexed *International Journal of Educational Research Open* (ScienceDirect), and a reviewer for the *International Journal of Police Science* (IJPS), the *International Journal of Research and Innovation in Social Science* (IJRISS), and BP International (India).

Dr. Mamdooh Abdelhameed Abdelmottlep is a former law enforcement officer with extensive professional experience in police training, security education, and international academic cooperation. In addition to founding the International Police Science Association (IPSA), Dr. Mamdooh created the World Internal Security and Police Index (WISPI) and Global Police and Security Index (GPSI), which measure the capacity and efficiency of police and security service providers to address internal security issues. Dr. Mamdooh's work has earned him numerous professional honors, including the 2014 CRIMINOLOGIA.IT, awarded by the United Nations, and the 2016 Distinction Medal, awarded by Sharjah's Ruler, His Highness Sheikh Dr. Sultan Al-Qasimi, for the continuous development of Middle Eastern Police Science and International Security Education. His research focuses on police science, law enforcement, and criminal justice administration.

Dr. Hector R. Garcia is an Associate Dean of the Graduate Criminal Justice Program at Southern New Hampshire University's Global Campus. He holds an Ed.D. in Organizational Leadership from Argosy University and multiple degrees in criminal justice and project management. A former Chief of Police and Dean at Miami Dade College's School of Justice, Dr. Garcia has held prominent leadership roles in law enforcement and academia. He has authored numerous scholarly works and has presented nationally for the FBI National Academy Associates (FBINAA), the U.S. Department of Justice, and the U.S. Department of Homeland Security. He has also served as a grant reviewer for NIH, NIJ, and OJJDP.

Chapter 2
Environmental Crime Scene Investigation *Processes and Practices*

Lakshmi Priya, Bindu Achathuparambil Gopalakrishnan, and George Kuttiparichel Varghese

1 Introduction

Crime, in its broadest sense, encompasses conduct that violates laws established by governing authorities and is punishable through legal sanction. Among the many emerging threats to global stability, environmental crime has gained increasing prominence due to its scale, complexity, and far-reaching consequences. Defined by the United Nations and INTERPOL as a collective term for illegal activities that harm the environment and benefit individuals, groups, or corporations—through exploitation, damage, trade, or theft of natural resources—environmental crime now ranks as the fourth largest criminal activity globally (United Nations Environmental Program, 2016). Growing at an alarming rate of 5% to 7% annually, environmental crimes generate an estimated 280 billion USD in profits each year, making them one of the most lucrative criminal enterprises worldwide (United Nations Environmental Program, 2016). These crimes manifest in diverse forms, such as illegal discharge of pollutants, oil spills, illegal wildlife trade, and the unauthorized trade of ozone-depleting substances. Their impact often transcends national borders, with pollution in one country affecting air and water quality in neighboring regions (Brisman & South, 2018).

L. Priya
Department of Civil Engineering, SCMS School of Engineering and Technology, Karukutty, Kerala, India

B. A. Gopalakrishnan
Department of Civil Engineering, Govertment Engineering College Thrissur, Thrissur, Kerala, India

G. K. Varghese (✉)
Department of Civil Engineering, National Institute of Technology Calicut, Kozhikode, Kerala, India
e-mail: gkv@nitc.ac.in

M. Mliless et al. (eds.), *Frontiers of Environmental Law Enforcement*,
https://doi.org/10.1007/978-3-032-10205-8_2

Contrary to popular perception, environmental crimes are not victimless. They contribute to species extinction, economic losses, disruption of livelihoods, climate instability, and public health hazards. Despite their widespread impact, these crimes are often overlooked or under-prioritized due to the complexities in tracing their consequences, thereby hindering effective legal and enforcement responses (Skinnider 2013). The United Nations General Assembly now recognizes environmental crime as a form of transnational organized crime, underscoring the increasing sophistication and coordination behind such activities. A survey conducted by the United Nations Environment Programme (UNEP) found that over 80% of countries acknowledge a significant correlation between environmental crimes and traditional criminal activities, including corruption (42%), counterfeiting (39%), drug trafficking (36%), cybercrime (23%), and financial crimes (17%). Alarmingly, the financial damage caused by environmental crime is estimated to be nearly 10,000 times greater than the resources allocated to combat it.

To confront this growing threat, international agencies such as INTERPOL and UNEP have developed specialized frameworks for addressing environmental crimes. However, the intricacies of these offenses demand scientific, structured, and interdisciplinary investigative approaches. The emerging field of Environmental Crime Scene Investigation (E-CSI) aims to bridge this gap by applying scientific methods and procedural rigor to uncover, document, and prosecute environmental offenses effectively (Priya et al. 2023).

This chapter explores the principles, methods, and challenges associated with E-CSI and highlights its vital role in ensuring environmental justice and sustainable governance.

2 Crime Scene Investigation in Traditional Crimes

The cornerstone of any successful crime investigation is the systematic collection and preservation of evidence (Lee 2020). In traditional criminal investigations, this process is initiated and led by law enforcement agencies, typically following the report of an offense—whether by a victim, witness, or via police surveillance. Once a potential crime is identified, the first responders are tasked with securing the crime scene, thereby preserving its integrity for further examination. At the core of crime scene investigation is the detection, collection, documentation, and analysis of evidence. This includes tangible physical evidence such as fingerprints, biological samples, and weapons, as well as digital or documentary evidence in the case of financial or cybercrimes. To reconstruct the sequence of events, investigators also rely on interviews with suspects, witnesses, and victims.

Upon arriving at the crime scene, officers approach with caution, taking note of all initial observations. The perimeter is secured, typically using police tapes, to restrict unauthorized entry and prevent contamination of the scene. Safety of the officers and any individuals present is a priority; any hazardous conditions must be addressed immediately, and medical assistance provided where necessary. Once the scene is deemed secure, the documentation process begins. This involves photographing, sketching, and recording detailed notes of the scene and evidence. All items of interest must be properly identified, collected, packaged, labeled, and secured to ensure the integrity of the chain of custody. A preliminary report is prepared summarizing the initial findings for use in further investigative procedures.

Forensic science has become an integral part of modern crime investigations, offering specialized tools such as DNA analysis, ballistics, toxicology, and digital forensics to enhance the reliability and objectivity of evidence. Emerging technologies, including surveillance systems, data analytics, and biometric identification, further strengthen investigative capabilities. However, the growing use of such technologies also raises important ethical and legal concerns, particularly with regard to privacy, civil liberties, and data protection.

An understanding of the principles and practices of traditional crime scene investigation provides a valuable foundation for examining how similar investigative rigor can be adapted to environmental crimes. Although environmental crimes differ in context and complexity, they similarly require a methodical, science-based approach to evidence collection, documentation, and analysis. Drawing parallels between traditional and environmental crime investigations highlights the importance of structured procedures, inter-agency coordination, and the application of forensic tools in ensuring effective enforcement and accountability.

3 Environmental Crime Scene Investigation

Environmental crimes, such as illegal waste disposal, chemical spills, unauthorized emissions, and illegal mining, pose unique and growing challenges to forensic investigation. Unlike traditional crimes, these offenses often involve diffuse sources, are spatially widespread, and may unfold over extended periods, making detection and investigation significantly more complex. As a result, E-CSI requires a tailored approach that integrates scientific expertise, interdisciplinary coordination, and legal clarity to effectively trace environmental harm and support enforcement actions.

While the general procedural framework for investigating environmental crimes shares similarities with traditional crime investigations—such as the need for systematic evidence collection, scene documentation, and analytical rigor—there are fundamental differences in the nature of the crime scenes, types of evidence, investigative personnel, and the tools required. In traditional crimes, evidence commonly includes physical items such as weapons, bloodstains, personal belongings, eyewitness statements, and electronic materials like CCTV footage. In contrast, environmental crime investigations rely heavily on scientific evidence—including soil, air, and water samples, laboratory test results, environmental monitoring data, and documentation from audits or regulatory inspections (Uhm and Nijman 2020). The crime scene itself also differs significantly. Traditional crime scenes are generally localized—a home, a bank, or a street—allowing for a more confined and controlled investigation. Environmental crime scenes, however, may span large and complex areas such as rivers, forests, wetlands, or industrial zones, making them logistically and scientifically more challenging to investigate.

The detection and reporting mechanisms further differentiate the two. Traditional crimes often come to light through direct reporting by victims or eyewitnesses, whereas environmental crimes are typically discovered through environmental audits, routine monitoring, or complaints from the public or NGOs. As such, the investigative responsibility also differs—traditional crimes are usually handled by police or criminal investigators, while environmental crimes require specialized environmental forensic experts trained in scientific sampling, data interpretation, and regulatory frameworks. When there is a convergence of traditional crimes with environmental crimes, investigations are best conducted by a collaborative team comprising police officers and environmental forensic experts (Priya et al. 2023). The investigation of environmental crimes typically requires a longer duration than traditional crimes. While many traditional cases are resolved within a relatively short time frame, environmental crime investigations often extend over longer periods due to the need for comprehensive sampling, laboratory analysis, and detailed impact assessments across broad and often degraded landscapes.

Given these complexities, the field of Environmental Forensics has emerged as a critical discipline to support the investigation and legal attribution of environmental crimes. This specialized area integrates principles of environmental science, analytical chemistry, geospatial analysis, and legal standards to trace sources of contamination, determine the timing and extent of damage, and identify responsible parties with scientific precision.

3.1 *The Role of Environmental Forensics*

Environmental forensics involves the investigation of environmental crimes by applying scientific methods and procedures to reconstruct the history of environmental harm and attribute responsibility. This discipline provides a structured framework to collect, analyze, and interpret environmental evidence, ensuring it is suitable for legal proceedings. It not only aids in determining the origin and extent of contamination but also supports enforcement actions and regulatory decision-making (Pivato et al. 2020).

Environmental forensics utilizes a wide range of interdisciplinary tools and techniques drawn from chemistry, geology, biology, environmental engineering, and hydrology. Investigations often begin with a review of historical documents to reconstruct operational practices or trace industrial changes over time. This is supported by advanced scientific techniques such as chemical and mineralogical fingerprinting, compound-specific isotopic analysis, and congener pattern analysis, all of which help identify pollutant signatures and link them to specific sources (Murphy and Morrison 2014). The discipline also employs dendroecology, where tree ring analysis is used to identify and date pollution events based on changes in growth patterns and contaminant uptake. In more complex scenarios, multivariate receptor models are applied to apportion pollution among multiple sources (Alimi et al. 2003).

Geospatial technologies such as remote sensing, aerial photogrammetry, and Geographic Information Systems (GIS) are used to map and visualize the spread of contamination over time and space (Grip et al. 2000). When investigating pollution that migrates through environmental media, tools like contaminant transport modeling, degradation modeling, and inverse modeling techniques are essential. These help simulate the movement and transformation of pollutants in soil, water, and air. Other methods include degradation product ratio analysis, chirality analysis of chemical compounds, surrogate chemical analysis, and the chronological review of changes in raw materials or chemical processes.

The proper application of these tools is dependent on the systematic and appropriate collection of environmental samples. For instance, if dendroecology is to be employed, tree core samples must be taken from the affected area. Similarly, radioactive isotope analysis would require sediment cores from lake or river bottoms, while contaminant transport modeling would necessitate a detailed understanding of the groundwater profile and the collection of representative groundwater samples across the site. In this way, the selection of forensic tool determines the overall sampling and analytical approach.

Ultimately, the choice and deployment of forensic tools shape the entire process of investigating an environmental crime scene. Each method comes with distinct data needs, sampling protocols, and analytical pathways, collectively guiding the trajectory of the investigation and its outcomes.

3.2 *Scene vs Scenario*

In the context of environmental crimes—particularly pollution-related offenses—the term ‘scene’ may not sufficiently capture the complexity and spatiotemporal nature of the criminal activity. Recognizing this limitation, Gopalakrishnan et al. (2024) have proposed the use of the term ‘scenario’ in place of ‘scene’, advocating for the concept of Environmental Crime ‘Scenario’ Investigation rather than the conventional ‘Scene’ Investigation. This distinction is not merely semantic; it reflects a fundamental difference in how environmental crimes unfold and are investigated.

Traditionally, a crime scene is defined as the specific place, time, and set of circumstances in which a criminal act occurs. It represents a fixed point on the space-time continuum—a snapshot of a crime event that is often limited in spatial scale and duration. It is typically confined to a single location, such as a room, a building, or a street corner, where the crime reaches its culmination and from which physical evidence can be collected and analyzed. The crime scene serves as a critical point of reference in reconstructing the actions leading to and resulting from the offense. In contrast, pollution crimes and other environmental violations are frequently diffuse, continuous, and spread across both time and space. For example, a factory illegally emitting toxic gases may cause harm not at a single location but across an entire downwind region over days, weeks, or even years. Similarly, illegal discharges into water bodies or gradual soil contamination from hazardous waste disposal are not isolated events but rather persistent activities that generate complex environmental impacts over extended periods. In such cases, the concept of a scenario is more appropriate. A scenario refers to a temporal progression of multiple interrelated events or conditions, encompassing a broader set of interactions and consequences. It allows for a more holistic view of environmental crimes—linking emission sources, environmental pathways, exposure routes, and resultant impacts across various locations and timeframes. This approach better supports the multidimensional investigation required in environmental crimes, including spatially distributed sampling, long-term data analysis, and interdisciplinary interpretation.

Therefore, in addressing pollution crimes, the terminology shift from 'scene' to 'scenario' provides a more accurate and functional framework for investigation. It aligns with the scientific and legal complexities inherent in such cases and emphasizes the need for comprehensive, scenario-based analysis to establish accountability and support judicial processes.

4 Investigation Process

Environmental crime investigations are inherently complex due to the nature of the crimes they address—often dispersed across large geographical areas, involving delayed detection, and embedded within intricate ecological systems. Unlike traditional crimes that are typically confined to a specific location and time, environmental crimes may unfold gradually, affecting air, water, soil, and biotic components over extended periods. This technical and temporal complexity necessitates a thorough, structured, and scientifically grounded approach to investigation.

International organizations such as INTERPOL and EUROPOL have developed comprehensive guidelines to assist in the investigation of environmental crimes (EUROPOL 2025). Notably, INTERPOL has published detailed manuals to standardize environmental forensic investigation practices across its member states (INTERPOL, 2014). These frameworks emphasize a sequence of critical steps: preliminary investigation, investigation planning, evidence collection, scene processing and documentation, laboratory analysis, engagement with victims and witnesses, and finally, reporting and legal action. Together, these stages form the backbone of a successful E-CSI.

4.1 Preliminary Investigation

The process begins with the collection of initial information to assess whether an environmental crime is likely to have occurred. Triggers for such investigations may include public complaints, unusual environmental phenomena like discolored water or persistent odors, reports from regulatory agencies or environmental NGOs, or findings from routine audits and surveillance. The credibility and urgency of these inputs are evaluated by the concerned authorities.

A preliminary assessment follows, where investigators visit the suspected site to visually inspect and document initial observations. This stage also involves gathering data on the site's topography, historical usage, compliance records, meteorological conditions, and previous environmental quality assessments. The purpose is to determine the nature and magnitude of the potential harm, any immediate threats to health or ecosystems, and whether the issue warrants a full-scale investigation.

4.2 Planning

Once a crime is reasonably suspected, investigators move to a detailed planning stage. This involves defining the scope of the investigation, allocating human, financial, and technical resources, and constituting an interdisciplinary team with relevant expertise. Safety protocols are established, especially when hazardous substances might be involved. Investigators must also secure any necessary permissions to access the site and formulate a strategy for systematic and legally defensible evidence collection.

4.3 Physical Evidence Identification and Collection

The core of E-CSI lies in collecting physical evidence that can link the environmental harm to its source. This step may involve the identification and retrieval of pollutants, waste residues, or contaminated environmental media (soil, water, air). The INTERPOL manuals and various national statutes, such as Section 11 of India's Environment (Protection) Act, 1986 (Ministry of Environment and Forests, India, 1986), provide procedural guidelines for such sampling. Samples must be collected in a scientifically sound and legally acceptable manner. Investigators must be familiar with the latest advancements in sampling techniques, legal standards, and analytical requirements. In transboundary or internationally significant cases, adherence to globally recognized protocols is especially important. Given the potential risks involved—such as exposure to hazardous substances—appropriate health and safety measures must be observed.

Advanced forensic tools and technologies, including drone surveillance, remote sensing, portable monitoring devices, GIS mapping, and chemical fingerprinting, are deployed to assist in evidence gathering. Each environmental crime presents a unique set of conditions, requiring tailored sampling strategies. Sampling must consider the type, quantity, and distribution of the materials collected and ensure proper labeling, preservation, and chain of custody documentation. Photographic and video evidence is also gathered to support later analysis and legal proceedings.

4.4 Scene Processing and Documentation

Thorough documentation of the crime scene is critical to maintaining the credibility and admissibility of collected evidence. Investigators prepare detailed field notes, assign unique identifiers to all samples, and map out their collection points. Observations from witnesses, the condition of visible materials, and interactions with the environment are meticulously recorded. These records form a foundational reference for both scientific interpretation and legal argumentation.

4.5 Laboratory Analysis

Laboratory testing enables the identification and quantification of contaminants, determination of their origin, and estimation of the time of release. This phase involves a spectrum of techniques—from basic chemical tests to highly sophisticated analytical tools such as gas chromatography-mass spectrometry (GC-MS), high-performance liquid chromatography (HPLC), inductively coupled plasma mass spectrometry (ICP-MS), and isotope or chirality analysis.

Such analytical techniques are essential for chemical fingerprinting, congener pattern recognition, degradation product analysis, and compound-specific isotopic profiling. These results help to confirm source attribution and assess the extent of contamination. Mathematical modeling techniques are often employed to simulate the spatial and temporal progression of pollutants and provide further insights into the cause-effect relationship of the environmental harm.

4.6 Handling Victims, Witnesses, and Perpetrators

In environmental crimes, the identification and handling of victims, witnesses, and perpetrators require sensitivity and precision. Victims may not always be individual persons but entire communities or ecosystems affected over time. Witnesses and accused parties must be approached with care to ensure the accuracy and protection of their statements. The confidentiality of their accounts must be respected, and all testimonies must be formally documented.

In addition to criminal accountability, investigations often involve determining compensatory measures for victims and calculating the environmental and economic costs of remediation. Understanding the human and ecological impact is a necessary component for effective prosecution and post-crime restoration.

4.7 Reporting and Legal Actions

The final step in the investigation process is the preparation of a comprehensive report that summarizes all findings, from field observations and sampling records to laboratory analysis and legal interpretations. This report is submitted to the relevant judicial or regulatory bodies. It must be factually accurate, scientifically sound, and legally robust, as it forms the basis for prosecution, remediation orders, or further regulatory action.

Care must be taken to ensure the integrity of the data presented. Instances of data falsification or analytical error must be avoided through stringent verification protocols. The report should also provide clear evidence of legal violations, the results of forensic analysis, and recommendations for corrective or punitive measures.

Each stage of the investigation process is intricately linked and sequentially builds towards reconstructing the full picture of the environmental crime. The success of the investigation depends not only on the collection of reliable evidence but also on the ability to logically and scientifically recreate the chain of events leading to the environmental harm. In other words, the investigation should help in Environmental Crime Scenario Reconstruction (E-CSR), where investigators integrate all collected data to establish the timeline of events, identify causative actions, and determine the parties responsible for the offense.

5 Reconstructing the Crime Scenario

Crime Scenario Reconstruction (CSR) is a critical forensic methodology used to interpret and recreate the sequence of events that led to an environmental crime (Chisum and Turvey 2011) While Crime Scenario Investigation (CSI) focuses on identifying, documenting, and collecting physical evidence to establish what and where the crime occurred, CSR goes further to answer the more complex questions of how the crime happened, who was responsible, and why it occurred. It aims to logically connect the available evidence to form a coherent, scientifically grounded narrative that can stand up to judicial scrutiny (Morgan 2017).

E-CSR builds upon the foundation laid by the E-CSI, using the documented evidence and investigative records to reconstruct the environmental crime event. This process is crucial for identifying perpetrators, establishing intent (where relevant), and supporting the enforcement of environmental laws. Each piece of evidence—whether physical, testimonial, or documentary—is essential in forming a clear, sequential account of the incident. The key stages of E-CSR are as follows:

5.1 Data Collection

The first stage in E-CSR is the comprehensive gathering of all relevant information. This includes laboratory analysis results, environmental samples, satellite data, visual documentation from the scene, as well as supporting documents such as permits, compliance reports, and historical records. Testimonials from witnesses, victims, or even the accused can also be crucial. It is essential to ensure that all accessible and legally admissible evidence is compiled before advancing to interpretation and analysis. Incomplete or poorly collected data at this stage can severely hinder the accuracy of the reconstruction.

5.2 Conjecture

In this stage, the investigators develop an initial, tentative explanation of the environmental crime based on the collected evidence. This conjecture serves as a working theory and provides direction for further inquiry. It is important to remember that this is a provisional assumption—one that must remain open to refinement or rejection as the investigation progresses and more evidence becomes available. The goal is to outline a plausible scenario that begins to connect the dots between the scene, the pollutants, and the possible actors involved.

5.3 Hypothesis Formulation

Here, the preliminary conjecture is refined into a formal hypothesis. Using all available data, the reconstructionist develops a structured explanation addressing the key investigative questions: how, who, what, when, where, and why. The hypothesis should align with the scientific evidence and be capable of explaining the observed contamination or environmental harm. Even in jurisdictions where strict liability applies and proving criminal intent (mens rea) is not required, presenting a logically sound and scientifically substantiated hypothesis strengthens the prosecution's case and helps meet the high evidentiary standards often demanded in criminal litigation.

5.4 Testing of Hypothesis

The formulated hypothesis is then rigorously tested by forensic experts for internal consistency and compatibility with all available data. It must be able to account for laboratory results, scene observations, environmental modeling, and witness statements. At this stage, investigators also critically assess and eliminate alternative hypotheses, ensuring that the proposed explanation remains the most plausible. In some cases, additional or missing evidence is sought to further substantiate or refute the hypothesis. A scientifically tested hypothesis provides a solid foundation for drawing conclusions and advancing toward theory formulation.

5.5 Theory Formulation

When a hypothesis successfully accounts for all observations, stands up to scientific scrutiny, and no alternative explanations remain viable, it evolves into a theory. This theory represents the reconstructed environmental crime scenario—an evidence-based account of the events that caused the environmental harm. It identifies the chain of actions, actors involved, and the resultant impacts on the environment and public health. The reconstruction thus serves as a powerful tool in legal and regulatory proceedings, supporting the prosecution of offenders with a clear narrative grounded in science and law.

5.6 Presentation Before Court

An additional stage in E-CSR, absent in traditional CSR, as suggested by Gopalakrishnan et al. (2024), is the presentation of the reconstructed scenario. Here, the reconstructed scenario is visually and interactively presented to legal authorities using advanced tools such as 3D models, animations, virtual reality simulations, and GIS-based crime scene maps. These technologies help translate complex scientific findings into a comprehensible format for judges, juries, and other non-expert stakeholders, enhancing the persuasive power of the evidence and increasing the likelihood of successful prosecution.

E-CSR transforms fragmented, technical evidence gathered during the investigation into a coherent, scientifically sound, and legally defensible narrative of the environmental crime. Each stage—from data collection to hypothesis testing, theory formulation, and courtroom presentation—plays a vital role in constructing a comprehensive understanding of the incident. This systematic process enhances the credibility and effectiveness of the investigation, while ensuring accountability by clearly establishing the causes, consequences, and responsible parties involved in the environmental harm. However, the foundation for a reliable and impactful E-CSR is a well-executed E-CSI.

6 Challenges to Effective Investigation of Environmental Crime Scenes

Investigation of environmental crimes often faces unique and multifaceted challenges that distinguish it from conventional criminal investigations. These challenges stem not only from the complex nature of environmental harm itself but also from societal, legal, and institutional shortcomings that hinder timely detection, accurate evidence collection, and effective prosecution. The following are some of the key challenges that impact the efficacy of environmental crime scene investigations:

6.1 Nature of Environmental Evidence and Its Transformation Over Time

Environmental crimes are inherently dynamic. The evidence—pollutants in soil, water, or air—is highly susceptible to degradation, dispersion, or natural transformation due to weather conditions, microbial activity, or chemical processes. By the time investigations are initiated, the concentrations of contaminants may fall below detection limits or drop below thresholds that constitute a legal violation. This transient nature of evidence makes timely detection and prompt forensic sampling critical, yet often unachievable, resulting in weakened cases due to insufficient or inconclusive data.

6.2 Low Priority and Limited Resources

Environmental crimes often do not evoke immediate societal outrage or a visible threat, unlike violent or property crimes. Consequently, they are frequently deprioritized by law enforcement agencies and policymakers. This diminished importance results in limited allocation of resources—be it personnel, training, equipment, or funding—compromising the depth and quality of investigations. In many jurisdictions, the environmental crime units are understaffed or function as an extension of regulatory bodies rather than dedicated forensic teams.

6.3 Dependence on Scientific Evidence and Its Inherent Uncertainty

Environmental crime investigations rely heavily on scientific evidence, such as chemical analyses and modeling results. However, all scientific methods carry a margin of error, which can be a significant hurdle in legal systems where proof beyond reasonable doubt is the standard for conviction. The complexity of analytical methods, limitations in precision, and possibilities of alternate interpretations can be exploited by defense attorneys to cast doubt on the prosecution's case. This challenge is exacerbated when there are no eyewitnesses or direct human testimony to support the forensic findings.

6.4 Establishing Guilty Intent (Mens Rea)

One of the foundational principles in criminal law is establishing mens rea, or the guilty mind. However, proving intent in environmental crimes is particularly difficult, as many such offences are committed through negligence, ignorance, or systemic oversight rather than direct criminal intent. Although many jurisdictions now adopt the principle of strict liability—especially in pollution-related crimes—where proving intent is not required, the application of this principle is not universal and may vary across regions and case types.

6.5 Trans-Jurisdictional and Regulatory Inconsistencies

Environmental crimes frequently span national and international boundaries, especially in cases of illegal wildlife trade, e-waste dumping, marine pollution, and cross-border industrial emissions. Investigating such crimes involves navigating multiple legal systems, each with its own regulatory standards, definitions, enforcement protocols, and degrees of institutional capacity. This lack of harmonization and inter-agency cooperation across jurisdictions often leads to jurisdictional conflicts, delays, and ineffective enforcement.

6.6 Weak Legal and Institutional Frameworks

In many regions, especially in developing countries, the legal infrastructure for addressing environmental crimes remains weak or outdated. Ambiguities in legislation, lack of specialized environmental courts, poorly trained judiciary, and low penalties contribute to a climate of impunity. Even when crimes are detected, the prosecution may be delayed or derailed due to loopholes in procedural law, bureaucratic inertia, or corruption. Weak institutional capacity also means that environmental inspectors and forensic experts may lack the authority or means to conduct comprehensive investigations.

6.7 Challenges in Reporting Environmental Crimes

Unlike traditional crimes, environmental offences often unfold gradually and silently, without triggering immediate public alarm. The lack of dramatic or visible consequences leads to delayed recognition and reporting. Public awareness about environmental crimes—such as unauthorized deforestation, sand mining, illegal waste disposal—is often low, especially in rural and marginalized communities. Furthermore, these crimes are sometimes committed by influential actors like corporate entities, politically connected individuals, or even government agencies. This power imbalance deters whistleblowing due to fear of retaliation or disbelief in receiving justice.

Language barriers, digital illiteracy, and geographical isolation add further obstacles to effective reporting mechanisms. In remote or underserved areas, people may not know how, or to whom, to report an environmental violation. Even when reports are made, they may not be acted upon with urgency or seriousness, leading to erosion of public trust in enforcement agencies.

Overcoming the challenges in environmental crime scene investigations requires systemic reform, capacity-building, and increased public engagement. This includes enhancing forensic capabilities, strengthening legal frameworks, ensuring interagency and cross-border cooperation, and fostering a culture of environmental responsibility. Only through concerted efforts across scientific, legal, institutional, and community domains can environmental crime investigations become effective tools for justice, deterrence, and ecological preservation.

7 Conclusion

The alarming rise of environmental crimes across the globe demands a swift, coordinated, and scientifically robust response mechanism. At the heart of this mechanism lies the Environmental Crime Scene Investigation, the foundational step that sets the trajectory for the entire enforcement and justice process. If errors occur in the early stages, particularly in evidence collection or preservation, the integrity of the investigation is compromised, leading to flawed conclusions and the possible escape of offenders from legal accountability.

Given the complexity of environmental crimes, where evidence may degrade over time or be diffused across ecosystems, crime scene investigation must be approached with meticulous attention and expertise. Investigators must possess not only a sound understanding of the chemical and biological behavior of pollutants in various environmental matrices but also be well-versed in modern analytical procedures and environmental forensic tools. These tools significantly influence how evidence is identified, collected, and interpreted, making them indispensable for building strong, admissible cases.

Environmental crimes often transcend borders, involving shared air, water, or migratory ecosystems. This highlights the critical need for inter-agency and international collaboration, where enforcement bodies, forensic experts, scientists, and legal authorities must work in harmony. Guidelines and manuals issued by global agencies like INTERPOL serve as essential resources for standardizing practices and fostering cooperation across jurisdictions.

A carefully executed crime scene investigation feeds into a credible and comprehensive Crime Scenario Reconstruction, enabling the reconstruction of events, identification of responsible parties, and delivery of justice. In this way, E-CSI is not just a technical exercise, but a powerful tool in upholding environmental justice and ensuring that those who degrade our planet are held accountable.

References

Alimi, H., Ertel, T., & Schug, B. (2003). Fingerprinting of hydrocarbon fuel contaminants: Literature review. *Environmental Forensics*, 25–38. https://doi.org/10.1080/15275920303489

Brisman, A., & South, N. (2018). Green criminology and environmental crimes and harms. *Sociology Compass,* 1–12. https://doi.org/10.1111/soc4.12650

Chisum, W. J., & Turvey, B. E. (2011). A history of crime reconstruction. In *Crime reconstruction,* 1–17. (2nd ed.), Academic Press.

EUROPOL. (2025). *Environmental crime.* Retrieved from https://www.europol.europa.eu/.

Gopalakrishnan, B. A., Varghese, G. K., Pivato, A., & Lega, M. (2024). Pollution crime scenario reconstruction for effective environmental redress. *Environmental Forensics,* 1–16. https://doi.org/10.1080/15275922.2024.2431313

Grip, W. M., Grip R. W., & Morrison, R. D. (2000). Application of aerial photography and photogrammetry in environmental forensic investigations. *Environmental Forensics*, 121–129. https://doi.org/10.1006/enfo.2000.0014

INTERPOL. (2014). Pollution crime forensic investigation manual. In *INTERPOL environmental security sub-directorate* (pp. 1–184).

Lee, Y. H. (2020). How police policies and practices impact successful crime investigation: Factors that enable police departments to "clear" crimes. *Justice System Journal*, 37–62. https://doi.org/10.1080/0098261X.2020.1719246

Ministry of Environment and Forests, India. (1986). The Environment (Protection) Act, 1986.

Morgan, R. M. (2017). Conceptualising forensic science and forensic reconstruction. Part I: A conceptual model. *Science & Justice: Journal of the Forensic Science Society*, 455–459. https://doi.org/10.1016/j.scijus.2017.06.002

Murphy, B. L., & Morrison, R. D. (2014). *Introduction to environmental forensics.* Academic Press.

Pivato, A., Gwinnett, C., & Varghese, G. K. (2020). Environmental forensics. *Detritus*, 1–2. https://doi.org/10.31025/2611-4135/2020.14021

Priya, L., Samad, A. M., Varghese, G. K., & Shah, I. F. (2023). Integrating environmental forensics in criminal investigation: Needs and methods. *Environmental Forensics*, 1–9. https://doi.org/10.1080/15275922.2023.2297414

Skinnider, E. (2013). Effect, issues and challenges for victims of crimes that have a significant impact on the environment. In *Emerging forms of crime that have an impact on the environment: lessons learned,* 1–10.

United Nations Environmental Program. (2016). The rise of environmental crime – A growing threat to natural resources peace, development and security. In R. Henriksen, A. Kreilhuber,

D. Stewart, M. Kotsovou, P. Raxter, E. Mrema & S. Barrat (Eds.), *A UNEP INTERPOL rapid response assessment*, 1–108. Retrieved from https://www.grida.no/publications/344
van Uhm, D. P., & Nijman, R. C. C. (2020). The convergence of environmental crime with other serious crimes: Subtypes within the environmental crime continuum. *European Journal of Criminology*, 542–561. https://doi.org/10.1177/1477370820904585

Lakshmi Priya has her Bachelor's in Civil Engineering (2013), Master's in Environmental Engineering (2015), and PhD in Environmental Forensics (2023). Her research interests include environmental forensics, environmental risk assessment, policy making, etc. She has more than 20 publications in journals and conferences. Lakshmi is currently working as an Assistant Professor at SCMS School of Engineering and Technology, Karukutty, Kerala, India.

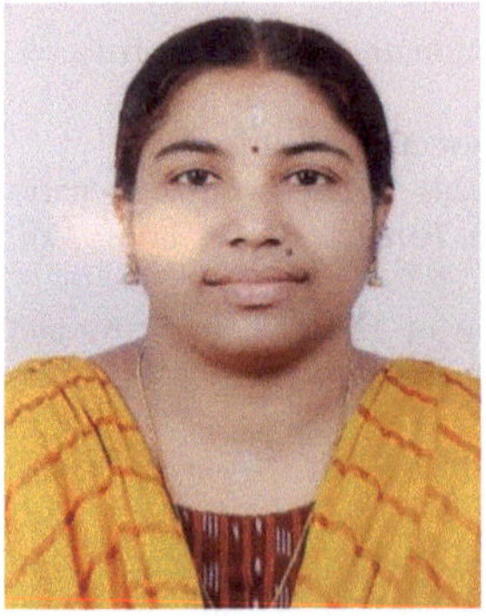

Bindu Achathuparambil Gopalakrishnan has completed her Bachelor's in civil engineering (1999) and Master's in Environmental Engineering (2015) and PhD in Environmental Forensics (2025). Bindu is currently working as an Assistant Professor at Government Engineering College Thrissur, Kerala, India. Her areas of research interest include environmental forensics and groundwater contamination. She has more than 15 publications in journals and conferences.

George Kuttiparichel Varghese is an Associate Professor of Civil Engineering at NIT Calicut, India, specializing in Environmental Forensics. He holds a PhD in the field, along with degrees in Civil and Environmental Engineering and a diploma in Environmental Law. He has guided eight PhD and 40 + postgraduate projects, and published over 130 scholarly works. He is an Associate Editor of Detritus and serves on national expert panels, including CPCB and MoEFCC. A consultant to government and industry, he is also a member of INEF, IWWG, and the Royal Society of Chemistry (MRSC).

Chapter 3
Tracing the Interrelation Between Organized Crime and Environmental Degradation: The Criminal Cartography of Environmental Destruction

Mapping Organized Crime's Role in Transnational Environmental Offenses

Hector R. Garcia

1 Introduction

Since the late 1990s, environmental crimes have exponentially grown and merged with organized crime, resulting in one of the top global security threats worldwide. Environmental crimes are well-planned and systematized and are routinely perpetrated by transnational organized crime (TOC) syndicates. TOC syndicates capitalize on weak regulations, institutional corruption, and technological anonymity to exploit and commodify the world's most vulnerable ecosystems (UNODC, 2024).

The illicit nexus of transnational organized crime with arms smuggling and drug dealing deeply disturbs biosystems while harming humanity, our society, and undermining human health and sociopolitical stability. UNEP (2023) reported that environmental crime has one of the most significant financial impacts, ranking as the fourth-largest global criminal enterprise, with annual revenue surpassing USD 281 billion. INTERPOL (2023a, 2023b) suggests that the resources gained during these crimes have a direct correlation to firearms, drugs, and human trafficking, creating significant challenges in the international police environmental and organized crime enforcement strata (Nellemann et al., 2020). The impact of this crime in nation-states with weak governance, especially in the Global South, yields pollution of water bodies, deterioration of soil, extinction of flora and fauna, and displacement of people,

H. R. Garcia (✉)
Southern New Hampshire University Global Campus, Criminal Justice Programs, Manchester, NH, USA
e-mail: hrgarcia@pm.me

M. Mliless et al. (eds.), *Frontiers of Environmental Law Enforcement*,
https://doi.org/10.1007/978-3-032-10205-8_3

which leads to increased global marginalization and inequities for economically disadvantaged populations (Dellas et. al., 2022; Bueger & Edmunds, 2023)

Researchers in environmental policy and green criminology have mapped what they call the Environmental Justice Lens into contemporary decision-making. The basic insight is straightforward but urgent: disadvantaged neighborhoods are often exposed to dirtier air, drink poorer water, and then discover they had no voice in the locations of landfills or factories. Those inequities echo across borders whenever transnational crime strikes. Illicit logging, clandestine waste dumping, and black-market wildlife trade hit Indigenous and low-income communities hardest because they lack political muscle and legal defenses. Analysts argue that bringing this perspective into compliance with audits and enforcement reports would nudge the system toward a fairer and more rights-driven reading of the law.

Emerging technologies are paving avenues for advanced artificial intelligence (AI) applications and smart predictive geospatial surveillance systems to be used in the fight against environmental organized crime; however, most countries still lack proper legal frameworks for the sophisticated law enforcement strategies needed to dismantle these fast-evolving criminal networks (Faure & Garoupa, 2009; Kriz, 2023). The existing green criminology literature primarily focuses on the environmental aspect of crime, paying little attention to the intersection of environmental crime with organized crime (Brisman & South, 2015; White, 2022). An alarming gap exists that has not synthesized organized crime and environmental crimes with advanced and emerging law enforcement investigative methodologies, such as AI-enabled detection and interdiction systems, predictive analytical programming, and geospatial surveillance (Faure & Garoupa, 2009; Kriz, 2023).

The goal of this chapter is to answer this oversight by providing insights into the complex frameworks of transnational organized crime and international environmental law. After analyzing the increasing role of technology as an innovator for providing solutions to these crimes, such as the use of AI for intelligent geospatial remote sensing, this work offers policies aimed at enhancing the existing fragmented structure of international environmental law enforcement systems.

2 Methods

This chapter employs qualitative and interdisciplinary approaches, utilizing document analysis, case study comparison, and environmental criminology theories. It focuses on understanding how organized crime groups intentionally engage in the wanton destruction of the ecology in various regions and across multiple forms of crime. This method facilitates the development of insights into the complexities of both governance gaps and operational strategies that enable these crimes to occur unchecked worldwide.

Data Collection and Sources

Primary data sources include institutional documents, assessments of transnational crimes, enforcement publications, and documents related to environmental matters from various international organizations. To this end, this research employs: 2020 to 2024 UNODC reports, UNEP's environmental crime assessments, INTERPOL's strategic briefings on environmental crime, CITES yearly enforcement figures, and relevant articles published in peer-reviewed journals on green criminology, environmental law, and security studies. Moreover, these findings were corroborated and contextualized with white papers produced by advocacy groups such as the Environmental Investigation Agency (EIA) and the Global Initiative Against Transnational Organized Crime (GI-TOC).

2.1 Case Selection Rationale

Three emblematic case studies representing the major domains of organized crime and environmental crimes formed the focus of a purposive sampling technique. Illegal logging in the Amazon Basin illustrates a corrupt nexus where forests are degraded through cartel-scale extraction activities. The e-waste dumping in West Africa reveals the covert movement of hazardous waste through maritime gaps made possible by front businesses. Trafficking of wildlife and endangered species through Southeast Asia focuses on the transportation facilities as a means for smuggling these precious commodities and facilitating financial crime. These cases are geo-strategically relevant and have been linked to organized crime groups, with reliable information shared across multiple agencies.

2.2 Analytical Framework

For guiding interpretation, this chapter employs a triangulated framework consisting of Convergence Theory, which focuses on how criminal businesses intertwine environmental crime with their other illegal activities, such as narcotics and human trafficking (UNODC, 2024). Secondly, the Environmental Security Paradigm frames these crimes as threats not only to ecological balance but also to global peace and order (Bueger & Edmunds, 2023). Thirdly, the Environmental Justice Lens examines the impact of these crimes on the under-governed regions that suffer the most due to inadequate governance, with vulnerable communities bearing the brunt of these crimes (Dellas et al., 2022).

Data were systematically categorized to identify trends in offenses, challenges faced by law enforcement, and technological innovations in policing. The analysis of each case study was contextualized socio-politically, focusing on identifying governance inadequacies and assessing the gaps in the application of surveillance and predictive analytics related to counter-crime measures.

3 Results

Transnational environmental crimes committed by TOCs exhibit unique characteristics, due to their regional origin. However, all share certain commonalities: lower risk and higher profits, minimal surveillance operations, and a low likelihood of being detected. TOC environmental criminal activity has yielded devastating consequences such as toxic pollution, deforestation, loss of native habitats, violence against environmental protection advocates, and government corruption, to name a few.

This chapter focuses on three case studies: illegal logging in the Amazon Basin, electronic waste dumping in West Africa, and wildlife smuggling through Southeast Asia, analyzing their operational infrastructures, socio-political repercussions, and technological responses. All cases illustrate a different type of environmental crime, collectively organized through a network exploitation of systemic gaps in weaker nations within the global environmental law enforcement system.

3.1 The Amazon Basin

Sustaining more than 60% of the world's remaining tropical rainforest, the Amazon Basin spans over nine countries and has become a focal point for environmental crime, particularly illegal logging. The scope and sophistication of these activities involve organized criminal collusion (UNODC, 2024; INTERPOL, 2023a). The lack of infrastructure and governance complicates matters in the Amazon, where organized crime groups exploit the region's remoteness, lax state presence, and weak rule of law to conduct timber piracy masquerading as legal silviculture through extensive logging and money laundering.

TOCs engage in large-scale timber piracy, primarily driven by the lack of documentation required to transit illicit wood to Europe and North America. Between 2018 and 2023, approximately 70% of logging in the Brazilian Amazon was attributed to illegal logging practices (Nellemann et al., 2020). These syndicates often legitimize timber shipments by using falsified documentation, shell companies, and bribing local officials. Public forensic accounting by INTERPOL and the World Bank estimates that international timber piracy results in the loss of over USD 10 billion in public revenue annually for the entire Amazon region (INTERPOL, 2023a).

Local narcotics rings and cartels, along with other criminal enterprises, have significantly increased their control over vast areas of the Amazon forest. These entities establish parallel government systems in rural towns, often extorting or forcibly displacing Indigenous peoples who attempt to resist deforestation (Human Rights

Watch, 2022). Moreover, environmental defenders in Brazil, Colombia, and Peru have encountered systematic aggression, with over 100 recorded murders of environmental activists each year in Latin America, many of which are linked to logging disputes (Global Witness, 2023).

Undoubtedly, the impact of corruption on illegal logging is significant. A 2022 report by Living on Earth noted active collusion between employees of regional forestry offices and timber traders in the Brazilian states of Pará and Rondônia. These arrangements facilitate the issuance of fraudulent harvesting licenses and protect the criminal offenders from criminal prosecution (Living on Earth, 2021).

Even with actions taken by international stakeholders, the enforcement of environmental crimes in this region remains weak. The FLEGT (Forest Law Enforcement, Governance, and Trade) and Brazil's Real-Time Deforestation Detection System (DETER) are primarily responsible for overseeing these areas and their natural resources. Local environmental law enforcement partners do not support these centralized entities, and recent political challenges have stripped many environmental protections from these zones, significantly impacting and virtually eliminating government environmental protections while reinforcing the status quo in these areas.

3.2 E-Waste Dumping in West Africa

Due to a lack of proper control and enforcement of policies, West Africa has now become a center for the illegal dumping of electronic waste (e-waste), with criminal syndicates emerging as the primary culprits. This region, particularly Ghana, Nigeria, and Côte d'Ivoire, receives tens to hundreds of thousands of metric tons of used electronics annually (Fig. 1). Although some of these devices are labeled as "second-hand goods," a considerable number of them are non-functional and should be treated as electronic waste (UNEP, 2023). Violations of the Basel Convention, which prohibit the export of hazardous waste from developed to developing countries without their prior permission, are widely facilitated using false documents, unsanctioned shipping routes, and inadequate customs enforcement (Secretariat of the Basel Convention, 2022).

Criminal groups establish connections with exporters in Europe, North America, and Asia to manipulate shipping manifests and mislabel boxes. By the time the waste reaches ports like Lagos or Accra, it is already integrated into the informal economy, which dismantles and incinerates e-waste without oversight, releasing lead, cadmium, mercury, and other heavy metals into the environment (Oeko Institut, 2010). One of the world's most notorious e-waste dump sites, Agbogbloshie in Ghana, typifies this environmental crime issue.

This has led to significant health effects for E-Waste workers in these West African countries (see Fig. 2). Research indicates that in Ghana, Nigeria, and Côte d'Ivoire (Fig. 3), unauthorized e-waste workers face significant health risks due to illegal burning and dismantling of e-waste. In all three countries, over 60% of

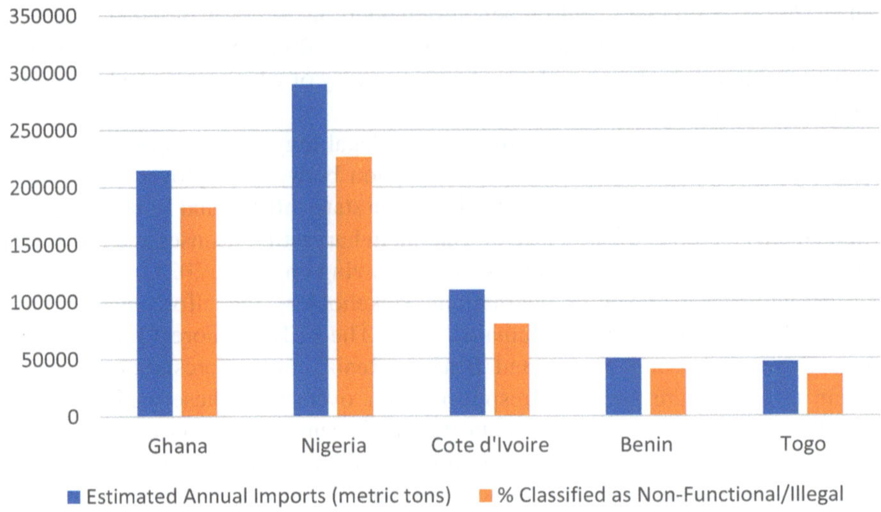

Fig. 1 E-Waste imports into West Africa (Retrieved from Unitar, 2024)

workers suffer from respiratory, neurological, and dermatological maladies that are attributed to excessive exposure to toxins such as lead and cadmium (Burns et al., 2019; Oeko Institut, 2010). This burden worsens the already dire health inequities faced by adolescents and children in these underprivileged communities, deepening the environmental injustice concerns associated with global e-waste dumping (Forti et al., 2020).

Subsequently, intermediaries control access to the site and extraction rights, informally taxing waste collectors and fostering a constrained micro-economy linked to transnational criminal networks (Forti et al., 2020). The toll on human health is devastating. Research conducted in Nigeria and Ghana has reported enriched neurotoxic profiles in children and chronic respiratory conditions among adult workers exposed to open burning (Burns et al., 2019). Given the absence of enforcement, underfunded frameworks, and corruption, policies like Nigeria's Extended Producer Responsibility (EPR) and Ghana's Hazardous and Electronic Waste Control Act are rendered ineffective.

The application of technology in surveillance remains rudimentary in the region. However, some pilot projects have demonstrated promise in tracing electronic waste (e-waste) shipments using blockchain technology, as well as employing mobile GIS for real-time tracking. The E-Waste Management in Ghana through Integrated Governance (E-MAGIN) initiative, supported by the European Union (EU), aims to foster public-private partnerships to enhance monitoring and reporting systems and is already raising awareness about the covert role of criminal facilitators (E-MAGIN Project, 2023).

This case illustrates the degree to which transnational organized crime strategically takes advantage of gaps in regulation between the Global North and South to

Fig. 2 *Note* Photo of West African E-Waste Worker (NPR, 2024)

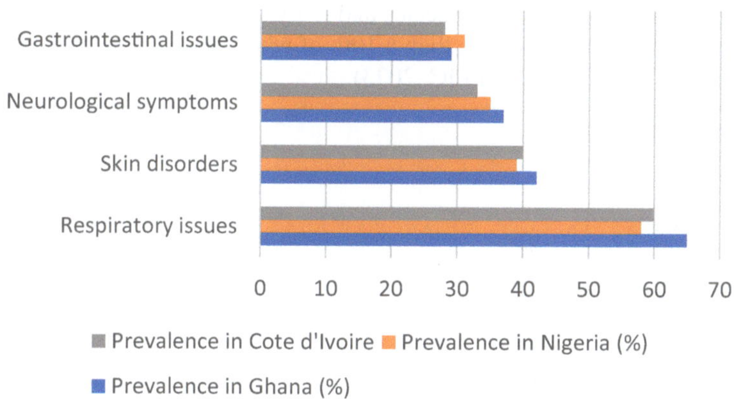

Fig. 3 West Africa E-Waste Workers Health Issues (Retrieved from: Burns et al., 2019; Oeko Institut, 2010)

outsource the toxic burdens of obsolete technology and transform West Africa into a dump site for the digital age.

3.3 *Wildlife Smuggling Through Southeast Asia*

The 11 countries that constitute Southeast Asia rank among the world's busiest corridors for illegal wildlife trade, combining the exportation and transportation of these natural commodities through international black markets. Criminal organizations exploit these Southeast Asian regions' sprawling ecosystems, targeting endangered species such as pangolins and particularly, their scales, and other exotic concoctions, serums, and souvenirs derived from tigers, rhinoceros, marine turtles, corals, ivory, and sharks. Criminal organizations easily traverse across borders that are thin on paper and thinner in practice (UNODC, 2024). Enforcement agents may try to impose the law, yet bribes, violence, and threats whittle enforcement back to a slow crawl, enabling the highly organized criminal networks to effortlessly move across international venues with ease.

Wildlife shipments often follow the same highways and waterways drug lords charted decades earlier, starting in Myanmar, Laos, or outer provinces of Indonesia before moving into and through Bangkok, Hanoi, or Kuala Lumpur and finally being shipped to buyers in southern China or farther afield (EIA, 2023). In a typical haul, smugglers will stack concealed shipments behind crates of farm produce, wrapping paperwork in forged bills of lading that even some port clerks no longer bother to check. During 2022, police in both Vietnam and Thailand (Fig. 4) seized approximately 23 metric tons of pangolin scales and ivory, a big headline yet, according to field estimates, probably no more than a 10th of what slipped away unrecorded the same year (CITES, 2023; TRAFFIC, 2023).

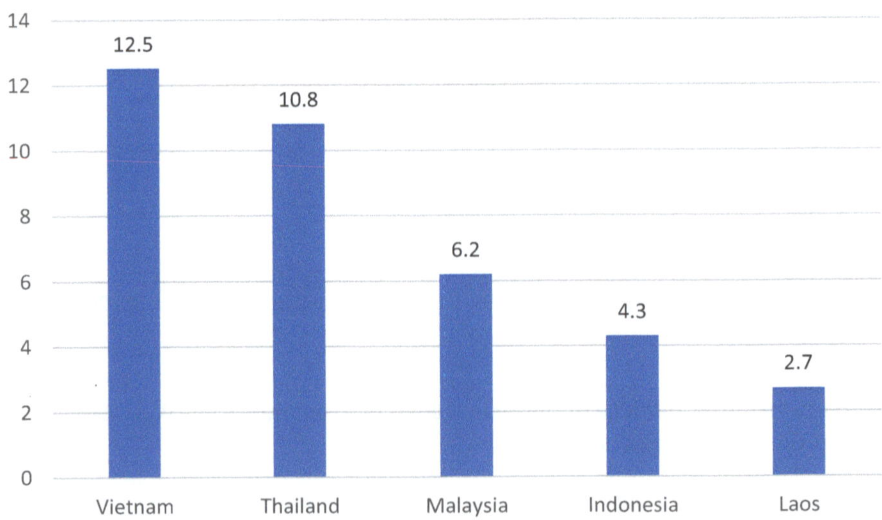

Fig. 4 Total wildlife contraband seizure (metric tons) (retrieved from: TRAFFIC, 2023: CITES, 2023; INTERPOL, 2023a)

Organized crime groups now treat wildlife trafficking as just another commodity, blending it with drugs and arms in their criminal enterprises. Chinese triads, Vietnamese syndicates, and African poaching rings regularly swap routes and contacts. The narco-leopard smugglers launder their profits in shell-import firms, high-end casinos, and condominiums bought in the name of distant shell companies (GI-TOC, 2022; UNODC, 2024). Corruption clouds the enforcement and response to these illegal enterprises. The ASEAN Wildlife Enforcement Network (ASEAN-WEN) has logged reports saying some customs and border officers in critical chokepoints engage in payoffs. The Convention on International Trade in The Endangered Species of Wild Fauna and Flora (CITES) is an international agreement designed to protect the survival of wild animals and species (CITES, 2025). CITES is active in its prevention efforts, yet the voluntary adherence to this international agreement takes a back seat to weak national penalties, giving traffickers room to conduct their illegal activities with little fear (WWF, 2022).

Countermeasures continue to mitigate the scope and scale of these criminal enterprises. In joint operations such as Operation Thunder and Operation Cobra, teams from ASEAN-WEN, INTERPOL, and UNODC have successfully seized large quantities of contraband and made numerous arrests. Advanced technologies such as DNA databases, social media venues used for contraband trade, and blockchain audits can track poachers from illegal contraband collection to its underground sales point. Unfortunately, rural districts stall because this technology is both costly and dependent upon reliable internet service, which is challenging, and even non-existent, in remote rural areas. (INTERPOL, 2023b).

This case depicts a stark picture: poachers and traffickers now run a supply chain as sleek and sophisticated as any Fortune 500 firm. Interrupting that flow demands more than arresting couriers; it calls for governments to shore up border law, merge wildlife statutes, and commit to designating real money behind forensics and persistent field surveillance.

4 Discussion

This chapter begins with an urgent question that looms larger every year: what happens when big-money crime meets big-picture ecology. Transnational syndicates are now hijacking forests, rivers, and even data streams, while inspectors and prosecutors still study the problem in isolation. The 24-h news cycle can headline "forest fires" or "corrupt officials," yet the harder story asks whether enough courtrooms, budgets, or bits of biometric software exist to dismantle a highly adaptive criminal enterprise that continuously adapts to mitigation and enforcement efforts. Weak rules once seemed like loopholes; they have started serving as deliberate launch pads for criminal enterprises that exploit system vulnerabilities.

Pulled from on-the-groundwork in three hemispheres, the evidence is arresting and, frankly, unnerving. Loggers with falsified permits fan out across the Amazon, e-waste hustlers personally escort their shipments through Lagos airport, and smugglers brazenly wave passports stamped in half a dozen capitals before boarding night flights from Kuala Lumpur. Though the cargo and the jungle differ, the playbook looks almost photocopied: front companies, bribed coast-guard captains, and ledgers that suddenly vanish the moment an investigator knocks. Experts at both the United Nations Office on Drugs and Crime (2024) and INTERPOL now label this entrenchment as emblematic of the 21st-century criminal enterprise, which is no longer a footnote in environmental law but its headline act.

Enforcement systems around transnational environmental crime still lag far behind the scale and sophistication of the offenses themselves. Although treaties such as CITES and the Basel Convention set useful, if aspirational, principles, their day-to-day application falters on issues of national prerogative, patchy funding, and overlapping legal authorities. High-tech fixes, such as satellite sweeps, blockchain cargo ledgers, and AI-powered watch lists, all look promising on paper, yet they stall without agreed protocols, timely data feeds, and teams trained to act across borders in real time (Faure & Garoupa, 2009; Kriz, 2023).

Drawing on Convergence Theory, the Environmental Security Paradigm, and core principles of Environmental Justice, this study sketches a multi-faceted lens for analyzing how organized crime leverages ecological fragility. When criminal networks operate in regions already weakened by environmental stress, they often hijack local governance structures and, in doing so, inflict the heaviest toll on the poorest households. Multidimensional thinking can therefore illuminate both the opportunistic mechanics of these crimes and their uneven human impacts.

Convergence Theory, as applied to environmental crime, looks at the way different illicit businesses operationally blend over time. Instead of treating arms deals, drug runs, and ecological violations as neatly separate cases, the model shows that various organized groups are crossing market borders on a single ledger. Take a hypothetical syndicate that hauls illegal teak one month and smuggles ivory the next; it often books both shipments onto the same freighter and pays off the same port foreman. By sharing financial tunnels and shipping lanes, the groups hedge their bets and slip past agencies that are still wired to compartmentalize criminal data. The strategy both conceals valuable cargo and exploits the enforcement gaps left when regulators focus on one offense at a time.

Contemporary environmental crime sits squarely at the nexus of organized illicit activity. Cartels no longer treat poaching or illegal logging as hobby supplements; those offenses sit shoulder to shoulder with cocaine, guns, and human trafficking in an updated ledger's core column. A single operation might see a lumber crew double as a money launderer by flipping timber cash into waterfront condominiums, while dockside e-waste brokers swap circuit boards for forged documents that finance a migrant smuggling trip. Through this lens, enforcement agencies are pressed to merge their environmental, economic, and maritime units into one task force rather than sending each category down a silo.

In practice, cartel operatives skim off carbon credits, trafficking syndicates hijack recycling chains, and the ecological bills for these maneuvers get sent to poorly governed regions. Displaced Indigenous peoples in the Amazon, children handling e-waste in coastal West Africa, and communities caught between corrupt wildlife rangers and poachers in Southeast Asia all share a debt they never incurred, a debt that arrives in the form of pollution, dispossession, or violence (Bueger & Edmunds, 2023; Dellas et al., 2022).

The Environmental Security Paradigm flips the script by treating ecological harm as an outright danger to national and international stability, not merely as a conservation side issue. Criminal syndicates or even state contractors can poison rivers, strip forests, or dump toxins in ways that rip apart the very fabric of local authority. Scarce clean water or disappearing game lands then fuels quarrels between neighbors, forces families to pack up and move, and deepens existing political rifts. From this vantage point, the health of watersheds, the continuity of forest cover, and the survival of key species stand as pillars of peace and effective governance (Bueger & Edmunds, 2023; Dellas et al., 2022).

The Environmental Justice Lens invites scholars to rethink who suffers first and most when land, air, or water is put at risk. It asks whether protection measures-land cleanups, safety nets, even park space-are filtered fairly or reserved for those with political influence. The framework draws on mid-century U.S. civil-rights activism yet has been taken up and adapted by organizers all over the world. Importantly, the people who suffer the most, tenant mothers next to toxic factories, tribal elders losing ancestral grazing, and low-income workers in export zone, often had the smallest hand in creating the hazards. Those burdens appear in daily seizures of breath, in quiet land takeovers, and in louder acts of police violence that follow public protest (Pellow, 2018; Schlosberg, 2007).

A second layer of the discussion notes that environmental crime, legal loopholes, and global supply chains share an appetite for cheap labor and loose oversight. Illegal mines gouged into Amazon headwaters spill mercury that creeps onto fields claimed by Indigenous families who never signed a contract. In coastal West Africa, teenage workers sift burned circuit boards on mud barges, breathing fumes that North American recyclers gladly ignore. Poachers then stroll into those same rural areas, stripping charismatic species while breaking any fragile tourism economy rooted in watching rather than shooting the animals. If policymakers want lasting fixes, activists argue, they cannot hand out band-aids alone; power must shift by letting affected communities set the terms of cleanup budgets, oversee enforcement officers, and draft the very laws meant to hold malefactors accountable.

Environmental justice probes the power dynamics that determine which groups are allowed to label an act a crime and which remain silent under its weight. Multinational agencies, for their part, tend to elevate trade embargoes or high-tech fixes. However, the people living on the poisoned ground usually demand land back, new ways to earn a living, and safety from those who might strike back after an exposure. Through this lens, the chapter argues that genuine progress in enforcement is not only evident in court verdicts but also in governance that repairs harm, invites everyone to the table, and treats nature as a shared inheritance rather than a bargaining chip.

Environmental crime is not simply a matter of shattered ecosystems; it generates chronic strategic headaches. When loggers slash forests in the Amazon or the Mekong, entire Indigenous communities are uprooted, and national borders blur.

In West Africa, heaps of discarded circuit boards leach toxins into groundwater and drain fertility from nearby fields. The long-term fallout erodes the legitimacy of local governments, leaving room for insurgents, organized gangs, or civil unrest to tighten their grip. Viewing these issues through the lens of environmental security pushes police and military planners to place habitat safeguards alongside human safety and national defense. That shift, in turn, forces agencies to pair park rangers with diplomats, intelligence analysts, and even military staff-something most environmental ministries never imagined doing.

Recent scholarship has often portrayed environmental wrongdoing as a swift, opportunistic pursuit of timber, ore, or waste-market profits. This chapter challenges that narrative by demonstrating how resource deterioration is now being orchestrated by sophisticated, tech-savvy crime syndicates that operate seamlessly across borders and sectors. In short, what once seemed a narrow ecological issue has now led researchers to point to a complicated policy conundrum, where authorities must juggle efforts against organized crime, the shifting landscape of global legal rules, and the push for truly sustainable development.

A meaningful policy response cannot stop seizing equipment or levying fines. It needs to shore up anti-corruption firewalls, align statutes from one country to the next, integrate technology and AI, weave data tools into everyday police practice, and elevate the voices of locals who bear the heaviest burdens. Absent a proactive regulatory framework, many law enforcement agencies will remain a step behind the agile criminal groups that already use the latest technological advances to subvert compliance regimes and skim profits from the green-energy economy.

5 Conclusion

Transnational organized crime has moved from the fringes of the environmental debate to center stage; it now designs many of the systems that strip land and ocean of their resilience. Fieldwork on three different continents shows that the typical white-collar, opportunistic wildlife seizure has hardened into routinized, well-funded commerce embedded deep within the global economy. Greedy operators probe and exploit every yawning loophole in customs law or mineral regulation while port-side brokers mask the ecological toll, leaving under-resourced communities to absorb the damage. Researchers describe a new phenomenon, calling it criminal ecology: an arena where ecosystems are logged or dredged like assets on a balance sheet and where would-be reformers suddenly discover their evidence has gone missing. Policymakers can no longer pigeonhole these violations as mere environmental infractions; they are entwined with national security, rule of law, and social justice. The way forward demands task forces that share data in real time, ground-level coalitions that elevate local voices, and audit-driven oversight that tracks dirty money from the jungle to the bank. Academics are urged to explore the intersection between blockchain forensics, anti-money laundering tactics, and human rights protocols, enabling law enforcement to stay one step ahead of syndicates that profit daily from a warming planet.

References

Brisman, A., & South, N. (2015). *Environmental crime and social conflict*. Routledge. https://www.routledge.com/Environmental-Crime-and-Social-Conflict-Contemporary-and-Emerging-Issues/Brisman-South-White/p/book/9781138637474?srsltid=AfmBOoqrGAS5Bx_rmDfMWv8g80Y757KGTC90pvtEo_mxw7pRDvb87_Xf

Bueger, C., & Edmunds, T. (2023). Blue crimes: Organized crime and environmental security in the maritime domain. *International Affairs, 99*(2), 493–510. Retrieved from https://www.sciencedirect.com/science/article/pii/S0308597X20300270

Burns, K. N., Sayler, S. K., & Neitzel, R. L. (2019). Stress, health, noise exposures, and injuries among electronic waste recycling workers in Ghana. *Journal of Occupational Medicine and Toxicology, 14*, 1.

CITES. (2023). *Annual illegal wildlife trade seizure data: Southeast Asia focus*. Retrieved from https://cites.org/eng/iata_cites_mou

CITES. (2025). *What is cites?* https://cites.org/eng/disc/what.php

Dellas, E., Pattberg, P., & Zelli, F. (2022). Global governance of environmental crime: Mapping fragmentation and synergy. *Earth System Governance, 12*, 100137. Retrieved from https://www.sciencedirect.com/science/article/pii/S1364815212002241

EIA. (2023). *Exposing the trafficking networks: Organized crime and the illegal wildlife trade in Asia*. Environmental Investigation Agency. Retrieved from https://eia-international.org/

E-MAGIN Project. (2023). *E-waste management in Ghana (E-MAGIN Ghana): From cradle to grave*. Retrieved from https://www.eeas.europa.eu/sites/default/files/39._sd_e-waste_management_in_ghana.pdf

Faure, M., & Garoupa, N. (2009). Environmental crimes. *Michael Faure, 2009. Criminal Law and Economics, chapter 13*. Edward Elgar Publishing.

Forti, V., Baldé, C. P., Kuehr, R., & Bel, G. (2020). *The global E-waste monitor 2020: Quantities, flows and the circular economy potential*. United Nations University. Retrieved from https://www.globalewaste.org/

GI-TOC. (2022). *Crime convergence and environmental trafficking in Asia-Pacific*. Global Initiative Against Transnational Organized Crime. Retrieved from https://globalinitiative.net/

Global Witness. (2023). *Decade of defiance: Ten years of reporting land and environmental activism*. Retrieved from https://globalwitness.org/en/campaigns/land-and-environmental-defenders/decade-of-defiance/

Human Rights Watch. (2022). *Rainforest mafias: How violence and impunity fuel deforestation in Brazil's Amazon*. Retrieved from https://www.hrw.org/report/2019/09/17/rainforest-mafias/how-violence-and-impunity-fuel-deforestation-brazils-amazon

INTERPOL. (2023a). *2,114 seizures of endangered animals and timber in major international law enforcement operation*. https://www.interpol.int/en/News-and-Events/News/2023/2-114-seizures-of-endangered-animals-and-timber-in-major-international-law-enforcement-operation

INTERPOL. (2023b). *Operation thunder: Coordinated global enforcement against wildlife crime*. Retrieved from https://www.interpol.int/en/News-and-Events/News/2023

INTERPOL. (2023). *Strategic report: Organized crime networks and environmental crime*. Retrieved from https://www.interpol.int/en/Crimes/Environmental-crime

Kriz, K. A. (2023). AI and the future of environmental law enforcement. *Environmental Law Reporter, 53*(7), 10625–10632. Retrieved from https://elr.info/news-analysis/53/10625/ai-

Living on Earth. (2021). *Corruption in Brazilian government*. Retrieved from https://www.loe.org/shows/segments.html?programID=21-P13-

Nellemann, C., Henriksen, R., Raxter, P., Ash, N., & Mrema, E. (Eds.). (2020). *Illegal trade in wildlife: The convergence of environmental crime with other serious crimes*. United Nations Environment Programme. Retrieved from https://wedocs.unep.org/handle/20.500.11822/32228

NPR. (2024). *Stunning photos of a vast e-waste dumping ground—and those who make a living off it. [Photograph]*. https://www.npr.org/sections/goats-and-soda/2024/10/05/g-s1-6411/electronics-public-health-waste-ghana-phones-computers

Oeko Institut. (2010). *Socio-economic assessment and feasibility study of sustainable e-waste management in ghana*. Oeko Institut. https://coilink.org/20.500.12592/6xf4jq

Pellow, D. N. (2018). *What is critical environmental justice?* Polity Press.

Schlosberg, D. (2007). *Defining environmental justice: Theories, movements, and nature*. Oxford University Press.

Secretariat of the Basel Convention. (2022). *Illegal traffic of hazardous wastes and other wastes: Guidance for national authorities*. Retrieved from https://www.basel.int/

TRAFFIC. (2023). *Illegal wildlife trade in Southeast Asia: Trends and enforcement challenges*. Retrieved from https://www.traffic.org/

Unitar. (2024). *The global e-waste monitor 2024-electoninc waste rising five times faster than documented e-waste recycling: UN*. https://ewastemonitor.info/the-global-e-waste-monitor-2024/

United Nations Environment Programme. (2024). *Environmental crimes are on the rise, so are efforts to prevent them*. https://www.unep.org/news-and-stories/story/environmental-crimes-are-rise-so-are-efforts-prevent-them

United Nations Office on Drugs and Crime. (n.d.). *Environmental crime: Trafficking in wildlifeand timber*. https://www.unodc.org/toc/en/crimes/environmental-crime.html

United Nations Office on Drugs and Crime. (2024). *Illegal wildlife trafficking is stubbornly prevalent, U.N. report says*. https://www.unodc.org/cofrb/uploads/documents/ECOS/World_Wildlife_Crime_Report_2024.pdf

UNEP. (2023). *State of knowledge of environmental crimes 2023: Global assessment report*. United Nations Environment Programme. Retrieved from https://www.unep.org/news-and-stories/story/environmental-crime

UNODC. (2024). *World Wildlife Crime Report 2024: Trafficking in protected species*. United Nations Office on Drugs and Crime. Retrieved from https://www.unodc.org/unodc/en/wildlife-and-forest-crime/index.html

White, R. (2022). *Advanced introduction to green criminology*. Edward Elgar Publishing. Retrieved from https://www.e-elgar.com/shop/gbp/advanced-introduction-to-green-criminology-9781800881086.html

WWF. (2022). *Corruption and illegal wildlife trade in the Greater Mekong*. World Wildlife Fund. Retrieved from https://origin-mekong.wwf-sites.org/challenges_in_the_greater_mekong/illegal_wildlife_trade2/

Dr. Hector R. Garcia is an Associate Dean in the Academic Strategy Management team at Southern New Hampshire University's Global Campus. He holds an Ed.D. in Organizational Leadership from Argosy University and multiple Master's degrees in criminal justice and project management. A former Chief of Police and Dean at Miami Dade College's School of Justice, Dr. Garcia has held prominent leadership roles in law enforcement and academia. He has authored numerous scholarly works and has presented nationally and internationally for the FBI National Academy Associates (FBINAA), the U.S. Department of Justice, and the U.S. Department of Homeland Security. He has also served as a grant reviewer for NIH, NIJ, and OJJDP.

Chapter 4
From the Statutory Basis to the Effectiveness of the Environmental Police's Mission in Morocco

Hind Majdoubi

1 Introduction

In February 2017, Morocco celebrated the inauguration of the first cohort of the Environmental Police, following up on the provisions of the framework law establishing the National Charter for the Environment and Sustainable Development, promulgated in 2014, which created the status of an Environmental Police. Indeed, Article 35 of the said framework law created the Environmental Police and entrusted it with the mission of strengthening the powers of the relevant administrations in matters of prevention, monitoring, and inspection.[1] The institutionalization of the Environmental Police is undoubtedly a major step forward in protecting the environment and combating harms that may affect it. However, it should be noted that the missions entrusted to it have existed in environmental laws promulgated since 2003. In fact, the decree governing the Environmental Police specifies that it is tasked with carrying out, in particular, the monitoring, recording, and ticketing (verbalization) provided for by Law 11-03 on the protection and enhancement of the environment, Law 12-03 on environmental impact assessments, Law 13-03 on combating air pollution, and Law 28-00 on waste management and disposal.[2] By way of illustration, the latter law states that its purpose is to prevent and protect human health, fauna and flora,

[1] Framework law establishing the National Charter for the Environment and Sustainable Development (2014), Article 35 (creation and mission of the Environmental Police.

[2] Full titles: Law 11-03 on the protection and enhancement of the environment; Law 12-03 on environmental impact assessments; Law 13-03 on combating air pollution; Law 28-00 on waste management and disposal.

H. Majdoubi (✉)
Laboratory for Studies and Research in Law and Political Science, Ibn Tofail University, Kenitra, Morocco
e-mail: hind.majdoubi@uit.ac.ma

Faculty of Legal, Economic and Social Sciences of Souissi, Rabat, Morocco

M. Mliless et al. (eds.), *Frontiers of Environmental Law Enforcement*,
https://doi.org/10.1007/978-3-032-10205-8_4

waters, air, soil, ecosystems, and the environment in general by establishing a system for controlling and sanctioning infractions committed in this field (Article 1). Thus, the principle of control clearly pre-dated the creation of the Environmental Police's status, but its effectiveness remained limited. On the one hand, the small number of officers authorized to record and ticket under the relevant administrations, combined with a lack of resources, made their missions almost invisible; on the other, judicial police officers, operating in a broader context of combating crime, conducted only very limited actions in the area of environmental protection.

Moreover, it must be emphasized that Morocco's environmental inspection and control system operates within a very complex institutional landscape involving various ministerial departments and supervisory bodies, all with roles in environmental protection.[3] In this context, the creation of the Environmental Police's status was a welcome development in Morocco. Nevertheless, eight years after its establishment, it is worth asking how effective it has been in protecting the environment. To that end, we will first present the status of the Environmental Police and, second, evaluate the progress achieved and the shortcomings to be addressed, with a view to improving the future legal framework.

2 Status of the Environmental Police in Morocco

In this section, we examine the Environmental Police's status from two angles: recruitment into the corps and, separately, its organization and mission.

2.1 Appointment of the Environmental Police

The Environmental Police is appointed by the government authority in charge of the environment—namely, the Department of Sustainable Development under the Ministry of Energy Transition and Sustainable Development. This hierarchical arrangement is salutary and avoids certain hierarchical drifts observed in other legal systems. In France, for example, Bergot characterized the environmental inspector as a ubiquitous figure who must exercise both technical and judicial competencies and answer to multiple authorities (Bergot, 2014).

The Department of Sustainable Development appoints officers assigned to the Environmental Police, under the title "inspectors of the Environmental Police," chosen from among the officers authorized under the environmental laws cited above. These officers may include civil servants, agents commissioned for this purpose by the competent administration or by local authorities, as well as sworn agents. To

[3] Ministry of Regional Planning, Water and the Environment; GTZ–PGPE; Water Management Consulting (2007). Study on the Environmental Inspection, Control and Surveillance System in Morocco, Phase 1, p. 25.

join the corps of Environmental Police inspectors, the order governing and specifying the operating procedures of the Environmental Police requires the cumulative satisfaction of three conditions.

Eligibility Criteria: Five Years' Service Requirement

This first condition clearly indicates that appointment to the Environmental Police requires proven training evidenced by a higher education degree obtained at least five years after the baccalaureate. Indeed, under the order governing the status of administrators, second-grade administrators must hold a degree requiring at least five years post-baccalaureate (Master's) or six years (Diplôme d'Études Supérieures Approfondies or Diplôme d'Études Supérieures Spécialisées: Advanced Higher Studies Diploma or Specialized Higher Studies Diploma). For the second category of eligible officers, candidates must hold a State Engineer degree in accordance with the order setting the status of engineers and architects.

This first condition underscores the requirement of advanced higher education to join the Environmental Police corps. It also reflects the substantial knowledge base expected of environmental police inspectors to conduct monitoring, investigations, and ticketing in environmental matters. That said, this first condition is complemented by a second.

Eligibility Criteria: Continuing Training Requirement

This continuing training aims to build Environmental Police officers' competencies in environmental prevention and protection, control and inspection techniques, and procedures for recording and ticketing infractions under environmental legislation and regulations. The modalities and curricula of this training are established by decision of the government authority responsible for the environment (to be specified by reference to the relevant decision).

This second condition operationalizes prospective members of the Environmental Police so they can carry out inspections and ticketing in strict compliance with the laws and their implementing texts, and master the required procedures. A lack of familiarity with the legal texts governing environmental protection, as well as the procedures required to establish liability, would undermine the Environmental Police's activities and could lead to inadmissibility of inspections and tickets. This is precisely the value of the second condition. In this regard, the Department of Sustainable Development has published a guide to environmental offenses, summarizing infractions and applicable sanctions, the administrative procedures to be followed, and measures to prevent, mitigate, or eliminate harm to the environment[4].

---

[4] Département du Développement Durable (Morocco). Guide des infractions environnementales [online guide summarizing offenses, procedures, and sanctions]. https://www.environnement.gov.ma/arabe/PDFs/Infractions.pdf.

Oath and ID Requirements for Inspectors

In addition to meeting the first two conditions, the candidate must take an oath in accordance with the legislation governing the oath for ticketing agents. Beyond the solemnity and symbolic weight of the oath, it is useful in two respects: first, it grounds the officer's responsibility to carry out public-service functions in compliance with legal norms; second, it establishes a presumption of veracity for statements reported orally or in writing in an official report (procès-verbal), which may only be rebutted by proof of falsity.

Once these conditions are met, those admitted to the Environmental Police corps, as inspectors, must identify themselves while on duty. To this end, they must prominently display a professional card evidencing their status. The card is issued in the format provided by the decree establishing the Environmental Police and delivered by the government authority in charge of the environment.

2.2 Organization and Missions of the Environmental Police

The Environmental Police reports to the government authority responsible for the environment, i.e., the Department of Sustainable Development. It should be nuanced, however, that the Environmental Police also exercises judicial police powers to record criminal offenses in the environmental domain. In this capacity, it reports to the Public Prosecutor's Office in accordance with Article 27 of the Code of Criminal Procedure[5]. Moreover, Environmental Police inspectors are placed under the authority of the administrations to which they belong and also under the authority of local authorities. In this respect, they perform administrative police functions.

More precisely, the decree on the organization and operating procedures of the Environmental Police clearly assigns two missions to the inspectors:

1. to carry out monitoring, inspection, research, investigation, recording of infractions, and ticketing as provided for in Laws 11-03, 12-03, 13-03, and 28-00;
2. to provide the necessary support to strengthen the powers of the administrations responsible for applying environmental-protection provisions contained in any special legislation.

First Mission

As previously indicated, the Environmental Police operates both preventively—against risks—and protectively—against harms that may affect the environment—under the aforementioned laws (11-03, 12-03, 13-03, and 28-00). As both an administrative police force and a body vested with judicial police powers, its first mission

[5] Moroccan Code of Criminal Procedure, Article 27 (reporting line of judicial police to the Public Prosecutor). https://www.menafatf.org/sites/default/files/Moroccan_AMLCFT_system_law_13_10.pdf.

encompasses monitoring, inspection, research, investigation, recording of infractions, and ticketing. As a judicial police force, it can conduct investigations and inquiries within the scope of criminal sanctions provided for by Laws 11-03, 12-03, 13-03, and 28-00. As an administrative police force, it monitors compliance by companies and individuals with applicable regulations—namely, conformity to authorizations granted and respect for established standards. Through inspections, the Environmental Police may identify violations or even risks of harm to the environment and require remedial measures.

To this end, inspectors are authorized to access installations and sites subject to control under the above-mentioned laws, and to carry out necessary checks, collect samples, and seize tools and items related to the infraction. In conducting inspections and recording infractions, they must respect the conditions laid down by the Code of Criminal Procedure.

Non-compliance with the applicable legal texts constituting infractions is recorded by the Environmental Police through the drafting of an official report (procès-verbal). This report must state, in particular, the circumstances and nature of the infraction and the explanations of the alleged offender. The website of the Department of Sustainable Development describes the reporting technique it adopts, whereby drafters summarize observed malfunctions, set out the measures recommended against the violator, and issue recommendations.

> Official reports thus drafted carry probative force until proven otherwise; in other words, a person contesting them must produce proof of falsity, notably as provided for in the Penal Code.
> Official reports must be transmitted by their authors to their administrations or to the courts, as the case may be, within deadlines ranging from 10 to 15 days depending on the violated legislative text.

Second Mission

The second mission—supporting administrations in applying special environmental-protection legislation—is detailed in Article 2 of the decree governing the Environmental Police. In practice, it entails: first, strengthening the capacities of agents authorized under special legislation to prevent, monitor, inspect, research, investigate, record infractions, and issue tickets in environmental matters, with particular attention to procedures and any needed supplementary training; second, pooling human and technical resources; third, standardizing work tools and ensuring effective information sharing; fourth, coordinating environmental control and inspection operations; and fifth, improving measures for prevention, risk forecasting, and pollution control.

Furthermore, the Environmental Police may act upon requests by any natural or legal person whose health or property has suffered harm due to the emission or discharge of a substance, sound, vibration, radiation, heat, or odor that has harmed their health or damaged their property. The victim has the right, within ninety days after the harm is established, to request the administration to conduct an inquiry.

The results of this inquiry must be communicated to the complainant. Under Law 11-03, in urgent cases the authority must provide information within a maximum of 60 days. Any refusal or closure of the case must be reasoned by the administration.

Law 11-03 requires that the results of the inquiry and any measures taken be notified to the applicant within 60 days. Similarly, under Law 13-03, any person who considers themselves harmed by an emission, release, or discharge of atmospheric pollutants and who requests the competent authority to conduct an inquiry must attach a medical or technical expert report to the request. In all cases, members of the Environmental Police are bound by professional secrecy and must not disclose data or information relating to projects subject to environmental impact assessment; any breach is punishable under the Penal Code.

Assessing the Environmental Police framework

The establishment of an Environmental Police marks a significant and beneficial advance in environmental monitoring and protection in Morocco. That said, improvements to the control system would be desirable.

2.3 *Positive Contributions of the Environmental Police*

Key advantages include strengthened environmental control and the development of environmental expertise among Environmental Police officers.

Strengthening Environmental Control

There is no doubt that the Environmental Police enhances environmental control and protection. This strengthening is apparent primarily at two levels. The first concerns monitoring compliance with the provisions of Laws 11-03, 12-03, 13-03, and 28-00 within the framework of administrative police work. The second concerns control within the framework of inquiries and investigations conducted as part of judicial police work.

At the administrative level, Environmental Police inspectors perform their missions on a surprise basis, at the request of the authority in charge of the environment, or within the framework of a National Environmental Control Plan. These inspections aim to verify the compliance of the activities of inspected parties with environmental standards in force and with the terms and conditions (cahiers des charges) attached to the authorizations granted.

As noted, such interventions may also occur under the National Environmental Control Plan, which the government authority in charge of the environment prepares annually, in consultation with the relevant authorities. Under Article 6, the plan pursues three objectives: first, to identify priority sectors and branches for environmental control—delineating control zones and selecting installations or activities based on environmental stakes and risk analysis; second, to set out a multi-year program to strengthen human and material capacities for supervising control and inspection operations; and third, to develop, harmonize, and simplify the tools and documentation used for research, investigation, recording, and ticketing of infractions under environmental protection legislation. Taken together, these measures clearly aim to strengthen environmental control by targeting priorities, optimizing control activities, and enhancing the effectiveness of the Environmental Police.

The second level concerns interventions by the Environmental Police in response to actual or suspected environmental harm. In such cases, the Environmental Police acts under its judicial police mandate and under the supervision of the Public Prosecutor's Office, supporting the judicial police in protecting environmental public order (Laville, 2015). Environmental protection is a public interest and a mission of maintaining public order, as the environment is a common good of the nation. During inspections, the Environmental Police may conduct judicial inquiries to gather evidence of violations that have caused harm to persons, property, or the environment.

Building on the Environmental Police's judicial mandate under the supervision of the Public Prosecutor's Office, recent national statistics (SNRTnews, 2024) clarify the scale and trajectory of enforcement. At a conference in Marrakesh on 17 July 2024, Mr El Hassan Daki, Prosecutor General of the King at the Court of Cassation and President of the Public Prosecutor's Office, reported that Moroccan courts registered 21,645 environmental cases involving 23,297 defendants in 2023, up from 19,575 cases and 20,362 defendants in 2022. This corresponds to increases of roughly 10.6% in cases and 14.4% in defendants, confirming intensifying judicial engagement against environmental crime. These trends underscore the centrality of prosecutorial oversight and the Environmental Police's referral role, and they reinforce the need for robust inter-agency coordination and transparent follow-up mechanisms outlined in the next section.

Finally, strengthening environmental control also occurs through cooperation—coordinating control and inspection operations and exchanging information on the one hand, and pooling human and technical resources on the other. Still, we must remember that effective environmental protection cannot be achieved by the Environmental Police and other specialized environmental police alone. Environmental associations (Flahault & Robic, 2007). and even citizens have important roles to play as guardians of the environment. Here, transparency and environmental communication are key to involving people by enabling access to reliable, complete, and relevant environmental information. In some legal systems, there is a growing call to extend the right of access to information—particularly environmental information—beyond public authorities to include data held by private parties, especially companies, while respecting trade secrets (Epstein, 2020). Consequently, environmental control by the Environmental Police would undoubtedly be reinforced by the participation of all stakeholders in this mission of public order, which is also a collective responsibility. However, such participation will only be effective if high-quality environmental information is accessible.

The Environmental Police and Environmental Expertise

Beyond the initial education, experience requirement, and successful completion of specialized training to join the Environmental Police, the corps also builds expertise through its mandate to strengthen the capacities of administrations covered by special environmental legislation. This capacity-building covers procedures to be followed, any necessary supplementary training, and improvements to prevention measures, risk forecasting, and pollution control.

Environmental Police expertise may also arise from acting upon requests by persons whose health or property has suffered harm due to environmental damage. Law 11-03, notably, allows any natural or legal person who has suffered harm due to an emission or discharge of a substance, sound, vibration, radiation, heat, or odor that harmed their health or damaged their property to request an administrative inquiry within 90 days of discovering the damage. A similar provision appears in Law 13-03 on combating air pollution, with the additional requirement that the applicant attach a medical or technical certificate. Law 13-03 provides that the results of the inquiry must be communicated to the applicant within sixty days, whereas Law 11-03 requires a 60-day response only in cases of urgency. In practice, these inquiries will likely be conducted by Environmental Police inspectors as part of their mission. Administrative responses setting out the results of these inquiries can, where environmental harm is established, later support damages claims by victims. Such administrative findings of violations and environmental harm—especially those arising from pollution—may provide victims with the necessary evidence to substantiate their claims. Resorting to private expert assessments can be costly, even prohibitive, for individuals or entities lacking the financial means to hire recognized experts.

2.4 *Shortcomings Calling for Desirable Improvements*

Environmental control by the Environmental Police faces certain limitations. First, it operates alongside other special police bodies, which calls for an overarching clarification of roles among all actors in environmental control. Second, disparities have been observed in how official reports (procès-verbaux) drafted by the Environmental Police are transmitted, which should be corrected. Finally, the Environmental Police's communication efforts should be improved in a spirit of transparency and continuous improvement.

Fragmentation of Control Over Environmental Harms

The creation of an Environmental Police in Morocco is unquestionably a major advance in control and protection, but these missions are also exercised under special laws by commissioned agents and civil servants for specific aspects of environmental protection. Examples include the water police; sworn and commissioned agents under the Coastal Law; and senior wardens and forest officers acting under the Dahir of 10 October 1917 on the conservation and exploitation of forests[6], among others. While the plurality of special police bodies may be read as reflecting the multiplicity of environmental dimensions, it has limits. Delhoste notes that this plurality can create complications, as each police authority must respect its own remit—limited by its object of intervention and by the specific procedures it must follow. As Lucidarme put it, "specialization without coordination can only ruin the action of Environmental Police" (Lucidarme, 2023).

Environmental control may be carried out by Environmental Police inspectors under the authority of the national environmental administration, but also by agents under the authority of local authorities commissioned for environmental control. If cooperation and pooling of resources and competencies among these controllers are not strengthened, effectiveness may be lost. Clarifying "who does what" would reasonably improve the Environmental Police's performance.

More broadly, codifying environmental law in Morocco would be desirable, clearly articulating the legislature's philosophy on environmental control and protection. A comprehensive Environmental Code would clarify whether the legislature favors amicable reparation of environmental harm before transmitting official reports to the courts, or whether it promotes the transaction mechanism provided for by Law 11-03. This also raises the question of how distinct or interdependent the various pieces of legislation should be. Although special laws pursue the ultimate aim of environmental protection, Law 11-03 presents itself as a general law aimed at protecting and enhancing the environment in all its elements. Clarifying the legislature's overarching approach to environmental protection therefore seems opportune.

Transmission of Official Reports Drafted by the Environmental Police

Differences can be observed in the follow-up to official reports drafted by the Environmental Police. Law 11-03 on the protection and enhancement of the environment—almost a framework law in its reach—does not clearly specify deadlines for transmitting official reports. It merely states that they must be sent "as soon as possible" to the competent court and to the governor of the relevant prefecture or province, subject to other legislative and regulatory provisions setting specific deadlines for prior administrative measures before initiating action—measures intended to put the offender on notice and require necessary remediation to eliminate environmental harm.

Under Law 12-03 on environmental impact assessment, the commissioned agent who records an infraction must draw up an official report and transmit a copy within no more than 15 days to the authority directly concerned with the project and another copy to the government authority responsible for the environment. After notification

[6] Dahir (Sherifian Decree) of 10 October 1917 on the conservation and exploitation of forests (Morocco).

to the concerned authority, the latter must put the offender on notice and invite them to comply with the law. If the offender refuses to comply and construction, development, or operation works are underway, the environmental authority, after notifying the relevant authority, transmits a copy of the report to the provincial or prefectural governor and to the president of the communal council to order a halt to the works pending the decision of the competent court. Law 13-03 specifies that official reports must be sent to the competent courts within 10 days from their drafting.

Under Law 28-00 on waste management and disposal, the administration has discretion: it may either serve a written notice requiring the offender to comply with legal and regulatory provisions, or transmit the official reports within fifteen days to the competent court when their conclusions warrant prosecution.

Due to these different processes, the Environmental Police must choose between friendly and confrontational methods, as well as different transmission deadlines and receivers, based on the applicable statute. Although the administration's preference for a peaceful resolution over litigation is understandable given some legislative goals, Moroccan law would benefit from reform that harmonizes the follow-up to official reports prepared by the Environmental Police in order to make it clearer and more effective.

Centralizing Information on Environmental Police Activities

The decree on the organization and operation of the Environmental Police contains very useful provisions on creating a database. This database is fed, on the one hand, by information collected from official reports on environmental matters, and on the other, by judicial decisions sanctioning environmental offenses.

Separately, the government authority responsible for the environment prepares, in consultation with the relevant government authorities and local government services responsible for environmental control and protection, an annual report on the Environmental Police's activities. This report must be prepared annually and sent to the Head of Government. It must then be published on the website of the authority in charge of the environment.

It is observed that the annual frequency reflects a desire to conduct short-term (one-year) periodic evaluations of Environmental Police activity. Also, reporting to the Head of Government indicates awareness—at the highest levels—of the importance of environmental protection. Finally, the obligation to inform the public also reflects a commitment to transparency. It is regrettable, however, that no data relating to the Environmental Police's activity has yet been published on the authority's website.

3 Conclusion

The establishment of an Environmental Police in Morocco is a welcome development. It acts to protect and enhance the environment, to ensure compliance by individuals and companies with precise legislative and regulatory provisions, and to act against those who violate the Law on Environmental Impact Assessments, the Law on Combating Air Pollution, and the Law on Waste Management and Disposal. However, eight years after celebrating the first Environmental Police cohort, there is a strong case for rethinking environmental control in Morocco—a system that is both fragmented and diluted due to overlaps between the Environmental Police and other special police forces for environmental protection. Progress will require codification of environmental law to provide a comprehensive vision for preservation and protection without unnecessary competition between corps and legal regimes. While the decree governing the Environmental Police laid foundations for cooperation and pooling of efforts and resources with other actors in environmental control, it would still be valuable for the legislature to provide a comprehensive clarification, including necessary specializations, to enhance effectiveness.

Moreover, any serious evaluation of environmental control in Morocco must consider the close relationship between the work of the Environmental Police and that of the judiciary. Transparent, complete, and regular communication by the environmental authority about Environmental Police activities is therefore highly desirable, to assess positive outcomes and address observed limitations and shortcomings.

Finally, confining environmental control to police bodies alone undermines the principle of collective responsibility for environmental protection, as advocated by Law 11-03 on the protection and enhancement of the environment. Engagement, participation, and awareness-raising for all stakeholders are thus the keystone of environmental preservation. Companies should be sensitized and encouraged to comply so as to avoid the need for intervention by the Environmental Police; in this respect, corporate compliance[7] warrants greater discussion in Morocco. The roles of associations and citizens in environmental protection in general—and environmental control in particular—are also well-established. All actors must therefore be sensitized, at multiple levels, to the stakes of their collective responsibility to preserve the environment as a legacy for future generations.

References

Bergot, D. (2014). L'inspection de l'environnement en pratique [L'inspection de l'environnement en pratique]. *Revue juridique de l'environnement, 39*(HS01), 171–176. Retrieved from https://droit-cairn-info.eressources.imist.ma/revue-juridique-de-l-environnement-2014-HS01-page-171?lang=fr

[7] Jombart, J. (2023). "Responsabilisation des entreprises, compliance et droit pénal." Revue juridique de l'environnement, 48(3), 527–536.

Delhoste, M.-F. (2001). *Les polices administratives spéciales et le principe d'indépendance des législations [Special administrative police forces and the principle of independence of legislation]* (pp. 37–46). LGDJ.

Département du Développement Durable. Contrôle environnemental [Environmental control]. Retrieved from https://environnement.gov.ma/fr/lois-et-reglementations/controle-environnemental?showall=&start=2

Département du Développemengt Durable. *Guide des infractions environnementales* [Guide to environmental offenses]. Retrieved from https://environnement.gov.ma/images/a_la_une/Publications%20PDF/dalil%2011%203(2).pdf

Epstein, A.-S. (2020). La compatibilité du droit d'accès aux informations environnementales de source privée avec la protection du secret des affaires: [The compatibility of the right of access to environmental information from private sources with the protection of business secrets]. *Revue juridique de l'environnement, spécial* (HS1), 137–149. Retrieved from https://droit-cairn-info.eressources.imist.ma/revue-juridique-de-l-environnement-2020-HS20-page-137?lang=fr.

Flahault, E., & Robic, P. (2007). Les associations d'environnement partenaires privilégiés des pouvoirs publics [environmental associations, privileged partners of public authorities]. In A. Dussuet & J.-M. Lauzanas (Eds.), *L'économie sociale entre informel et formel. Paradoxes et innovations.* (pp. 183–198). Presses universitaires de Rennes. ⟨halshs-00777029⟩

Jombart, J. (2023). Responsabilisation des entreprises, compliance et droit pénal [Corporate accountability, compliance and criminal law]. *Revue juridique de l'environnement, 48*(3), 527–536. Retrieved from https://droit-cairn-info.eressources.imist.ma/revue-juridique-de-l-environnement-2023-3-page-527?lang=fr

Laville, B. (2015). L'ordre public écologique: des troubles de voisinage à l'aventure de l'anthropocène [Ecological public order: from neighborhood disturbances to the adventure of the anthropocene]. *Archives de philosophie du droit, 58*(1), 317–336. Retrieved from https://doi.org/10.3917/apd.581.0342

Lucidarme, A. (2023). Une spécialisation sans coordination n'est que ruine de l'action des polices environnementales [specialization without coordination only ruins the action of environmental police.]. *Revue juridique de l'environnement, 48*(3), 537–546. Retrieved from https://droit-cairn-info.eressources.imist.ma/revue-juridique-de-l-environnement-2023-3-page-537?lang=fr

Ministère de l'Aménagement du Territoire (2007). *Étude sur le Système d'Inspection, de Contrôle et de Surveillance de l'Environnement au Maroc.* [Study on the environmental inspection, control and monitoring system in Morocco]. Retrieved from http://www.abhatoo.net.ma

SNRTnews. (2024, July 11). *Criminalité environnementale: plus de 23.000 personnes poursuivies au Maroc.* https://snrtnews.com/fr/article/criminalite-environnementale-plus-de-23000-personnes-poursuivies-au-maroc-99821

Hind Majdoubi is a tenured professor of law at Ibn Tofail University, Kenitra, Morocco, and an associate professor at the Faculty of Legal, Economic and Social Sciences of Souissi Rabat. She works mainly in the field of environmental protection and sustainable development. Her research has focused on environmental protection, the fight against waste and action against vulnerabilities and discrimination among other issues. In this context, she has made several communications and publications.

Chapter 5
Pedagogy for Teaching Environmental Crimes to Law Enforcement Officers

Human-Centric Strategies and Global Perspectives

Mamdooh Abdelhameed Abdelmottlep

1 Introduction

Environmental crime represents a growing threat to ecological stability, biodiversity, and human welfare. These offenses, ranging from illegal deforestation and wildlife trafficking to hazardous waste dumping, not only degrade ecosystems but also exacerbate socio-economic inequalities, disproportionately affecting marginalized communities (Lynch et al., 2018). Law enforcement agencies worldwide face significant challenges in combating these crimes because of their transnational nature, legal complexities, and the involvement of organized criminal networks (UNODC, 2021). Developing an effective pedagogical framework for training law enforcement officers requires a human-centric approach that integrates the principles of social justice, ethical enforcement, and community engagement. This chapter examines contemporary strategies for educating officers on environmental crimes, drawing on criminological theories, interdisciplinary methodologies, and global case studies. By synthesizing academic research with practical training models, this chapter provides a comprehensive guide for designing curricula that enhance officers' investigative, analytical, and ethical decision-making capacities.

M. A. Abdelmottlep (✉)
International Police Science Association, Katy, TX, USA
e-mail: drmamdooh@outlook.com

M. Mliless et al. (eds.), *Frontiers of Environmental Law Enforcement*,
https://doi.org/10.1007/978-3-032-10205-8_5

2 Human-Centric Foundations in Environmental Crime Pedagogy

2.1 Incorporating Social Justice and Ethical Principles

Environmental crimes do not occur in a vacuum; they are deeply embedded in systems of inequality, where marginalized communities—particularly Indigenous peoples, low-income populations, and developing nations—bear the brunt of ecological harm (White, 2014). These crimes often exacerbate existing socioeconomic disparities, as vulnerable groups lack the political and economic power to resist environmental exploitation. For example, illegal mining operations in the Democratic Republic of Congo have caused irreversible ecological damage and led to forced displacement, child labor, and violent conflicts over resource control (UNEP, 2020). Similarly, in the Amazon Basin, Indigenous communities face land invasions and deforestation, threatening their cultural survival and traditional ways of life (Brazilian Institute of Environment, 2020). These cases illustrate that environmental crimes are not merely violations of the law but also of human rights, demanding a justice-oriented approach to law enforcement training.

To address these systemic injustices, training programs must integrate ethical frameworks that guide officers to balance ecological protection with social equity. Environmental ethics, rooted in sustainability, intergenerational justice, and ecological rights, should be a core curriculum component (Elliott, 2016). Officers must be trained to recognize that enforcing environmental laws is not just about penalizing offenders but also about protecting communities from harm. For instance, in cases of illegal waste dumping in low-income neighborhoods, officers should be equipped to assess not only regulatory violations but also the public health consequences for residents of the area. Ethical decision-making models, such as consequentialist and deontological approaches, can help officers navigate complex scenarios where legal enforcement intersects with human welfare (Lynch et al., 2018).

Furthermore, community-centric approaches are essential for fostering trust and cooperation between law enforcement and affected populations. Traditional policing models often adopt a top-down enforcement strategy that can alienate communities that are already distrustful of authorities. Instead, training should emphasize participatory methods in which officers engage with local stakeholders to co-develop enforcement strategies. A notable example is Indonesia's community-based fisheries monitoring, where local fishers collaborate with law enforcement to report illegal fishing activities, leading to more effective and culturally sensitive interventions (Philippine National Police, 2021). Similarly, in Canada, Indigenous-led conservation initiatives have improved wildlife protection by incorporating traditional ecological knowledge into enforcement practices (Canadian Wildlife Federation, 2022). Such models demonstrate that when law enforcement agencies work *with* rather than *against* communities, compliance improves, and environmental crimes are more effectively deterred.

A critical aspect of social justice in environmental law enforcement is the recognition of environmental racism, where pollution and ecological degradation disproportionately affect minority and low-income communities. Cases such as the Flint water crisis in the United States highlight how systemic neglect and weak enforcement can lead to long-term public health disasters (Lynch et al., 2018). Therefore, training programs must include modules on environmental justice to teach officers to identify and challenge discriminatory enforcement practices. Role-playing exercises and case study analyses can help officers understand implicit biases and develop strategies for equitable law enforcement.

Finally, ethical training should extend to corporate and governmental accountability. Many environmental crimes are perpetrated by powerful entities, such as multinational corporations and corrupt officials, who exploit regulatory loopholes. Officers must be trained to recognize corporate malfeasance and political interference to ensure that enforcement actions are impartial and transparent. The inclusion of whistleblower protection protocols and anti-corruption measures in training curricula can empower officers to resist undue influence and uphold justice (UNODC 2021).

By embedding social justice and ethical principles into environmental crime training, law enforcement agencies can shift from a purely punitive approach to one that prioritizes human dignity, ecological integrity and restorative justice. This paradigm not only enhances enforcement effectiveness but also strengthens the relationship between officers and the communities they serve.

3 Highlighting Global Interconnections

Environmental crimes frequently operate across borders, necessitating an international perspective in training (INTERPOL, 2021). The transnational nature of offenses such as illegal wildlife trade, electronic waste dumping, and cross-border pollution requires law enforcement agencies to adopt a globally coordinated approach to combat them. Criminal networks exploit regulatory gaps between nations by leveraging differences in legal frameworks and enforcement capacities to evade prosecution. For example, the illegal trade in endangered species often involves source countries (where poaching occurs), transit hubs (where smuggling networks operate), and destination markets (where demand drives trade) (UNODC, 2021). Without international cooperation, efforts to dismantle these networks are fragmented and ineffective.

Comparative Case Studies play a crucial role in helping officers understand the complexities of global environmental crime. Analyzing transnational operations, such as Interpol's crackdown on wildlife trafficking syndicates, provides valuable insights into investigative techniques, intelligence-sharing mechanisms and jurisdictional challenges. Operations such as Thunderstorm (2018) and *Golden Strike* (2022) which targeted illegal wildlife trade across multiple continents, demonstrate how coordinated law enforcement efforts can disrupt criminal supply chains (Interpol, 2021). Training programs should incorporate these case studies to illustrate the best practices in international collaboration, emphasizing the importance of real-time data exchange and joint task forces.

Additionally, Cultural Sensitivity Training is essential to ensure that enforcement strategies respect local customs and indigenous rights. Many environmental crimes occur in regions where communities have strong cultural and spiritual ties to natural resources. For instance, India's sacred groves—forests protected by indigenous communities for religious and ecological reasons—are often targeted by illegal loggers. Officers trained in cultural awareness are better equipped to engage with local stakeholders, avoiding enforcement actions that could inadvertently criminalize traditional practices (Brazilian Institute for the Environment, 2020). Similarly, in the Amazon, recognizing Indigenous land rights has proven critical in combating illegal deforestation, as local communities serve as frontline defenders of their territories (IBAMA, 2020).

The role of international legal frameworks must also be emphasized in training programs. Treaties such as the Convention on International Trade in Endangered Species (CITES) and the Basel Convention on Hazardous Waste provide legal foundations for cross-border enforcement. However, officers must understand how these agreements are implemented at the national and local levels, where enforcement capacities may vary widely. Training should include simulations of multi-jurisdictional investigations, where officers navigate different legal systems and diplomatic protocols to build prosecutable cases (UNEP, 2020).

Finally, technology and data-sharing platforms are transforming global enforcement efforts. Tools such as the World Customs Organization's environmental crime database and INTERPOL's Global Fisheries Enforcement network enable real-time tracking of illicit activities. Officers should be trained to use these platforms to identify trafficking routes, monitor suspicious shipments, and collaborate with international counterparts (EU Environmental Enforcement Network, 2019). By integrating these global perspectives into training curricula, law enforcement agencies can better address the borderless nature of environmental crime and foster a more unified and effective response.

4 Expanding Pedagogical Approaches

4.1 Theoretical Foundations in Environmental Crime Education

Criminological theories provide critical insights into environmental offenses, shaping prevention and enforcement strategies by offering conceptual frameworks for understanding offender motivations and crime patterns. Routine Activity Theory (Clarke, 1997) has proven particularly valuable in urban environmental enforcement, as demonstrated by its application in Mexico City's monitoring of industrial zones that are prone to illegal waste disposal. By analyzing the convergence of motivated offenders, suitable targets, and the absence of capable guardianship, law enforcement agencies can develop targeted surveillance strategies for high-risk areas. This theoretical approach helps officers move beyond reactive policing to anticipate and prevent environmental violations before they happen.

Rational Choice Theory offers another valuable lens for understanding environmental crimes by examining the cost-benefit calculations offenders make. In South Africa, this theory has informed anti-rhino poaching tactics by analyzing how poachers weigh potential profits against the risks of detection and punishment (Kenya Wildlife Service, 2022). Training programs incorporating this perspective teach officers to disrupt criminal decision-making by increasing perceived risks through visible patrols while simultaneously decreasing potential rewards through market reduction strategies. Such theoretically grounded approaches have led to more sophisticated and effective enforcement methodologies in conservation areas worldwide.

Environmental Justice Theory (Lynch et al., 2018) provides the most socially relevant framework, highlighting systemic disparities in pollution exposure and environmental protection. The Flint water crisis serves as a stark case study that demonstrates how marginalized communities often bear disproportionate environmental burdens. Training programs that incorporate this theory help officers recognize patterns of environmental racism and develop enforcement strategies that prioritize vulnerable populations. By understanding the socio-political dimensions of environmental crime, officers can better address the root causes rather than just the symptoms of illegal activities.

5 Practical Skill Development for Field Readiness

Experiential learning methodologies bridge the gap between theoretical knowledge and practical application, creating law enforcement professionals capable of handling complex environmental-crime scenarios. Virtual Reality Simulations represent a cutting-edge approach to training, with Norway pioneering their use for illegal logging crime scene analysis (Canadian Wildlife Federation, 2022). These immersive technologies allow officers to practice evidence collection, chain-of-custody procedures, and forensic documentation in controlled but realistic environments prior to encountering actual crime scenes. The ability to recreate various scenarios, from wildlife crime scenes to chemical spill investigations, makes VR an invaluable tool for comprehensive training.

Field Exercises provide irreplaceable hands-on experience that classroom instruction cannot replicate. Kenyan ranger training programs in national parks exemplify this approach, where trainees learn to track poachers, identify trafficking routes, and collect forensic evidence in conservation areas. These exercises often involve multi-day simulations whichere trainees must navigate challenging terrain, interact with local communities, and make rapid enforcement decisions under realistic conditions. Such immersive training develops not only technical skills but also the judgment and adaptability required for effective environmental-law enforcement.

Diagnostic Assessments are critical tools for measuring training effectiveness and identifying knowledge gaps. Slovenia's implementation of pre- and post-training evaluations (Slovenia Ministry of Environment, 2022) demonstrates how systematic assessment can refine curriculum development and delivery. These evaluations often combine written examinations with practical skill demonstrations, ensuring that officers meet competency standards in both knowledge and application. Regular assessments also allow training programs to evolve in response to emerging environmental crime trends and new enforcement technologies.

6 Interdisciplinary Integration for Comprehensive Enforcement

The complex nature of environmental crime demands training approaches that transcend traditional disciplinary boundaries. Legal Training forms the foundation of effective enforcement, as demonstrated by Brazilian officers' study of the Forest Code to combat Amazon deforestation (IBAMA, 2020). Comprehensive legal education must cover not only domestic environmental statutes but also international treaties and conventions, as many environmental crimes have cross-border implications. Specialized training in evidentiary standards and courtroom procedures ensures that cases are properly developed for successful prosecutions.

Scientific Literacy has become increasingly crucial as environmental crimes become more technically sophisticated. Australian programs that teach hydrology to trace illegal pollutants (Australian EPA, 2022) exemplify how scientific training can enhance enforcement capabilities. Officers must understand the basic principles of ecology, chemistry, and environmental forensics to properly investigate crimes ranging from toxic dumping to wildlife trafficking. Partnerships with academic institutions and research organizations can provide access to specialized knowledge and laboratory resources that support these field investigations.

Economic Analysis offers valuable insights into the financial drivers of environmental crime. Nigerian training programs that examine the economic incentives for oil theft have led to more targeted and effective enforcement strategies. Understanding black market dynamics, supply chain vulnerabilities, and money laundering techniques enables law enforcement officers to disrupt criminal enterprises more effectively. This economic perspective also helps develop prevention strategies that address the root causes rather than just the symptoms of environmental crime.

The integration of these diverse disciplines creates law enforcement professionals who can address environmental crimes from multiple perspectives. By combining legal expertise with scientific knowledge and economic understanding, officers can develop more comprehensive strategies for prevention, investigation and prosecution. This interdisciplinary approach also fosters better collaboration with experts from other fields, creating more effective enforcement networks to combat complex environmental crime.

7 Collaborative and Digital Strategies in Environmental Crime Enforcement

7.1 Strengthening Multi-Agency Collaboration Frameworks

The complex nature of environmental crimes necessitates the breaking down of traditional silos between agencies and organizations. Multi-agency collaboration has emerged as a critical strategy, with training programs increasingly incorporating joint exercises that mirror real-world enforcement challenges to this end. In the Philippines, integrated training programs have brought together maritime police, marine biologists, and coastal community leaders to combat illegal fishing practices through scenario-based drills (Philippine National Police, 2021). These exercises simulate actual enforcement situations in which participants must coordinate surveillance operations, evidence collection, and community engagement strategies. Such training not only builds technical competencies but also fosters a mutual understanding of each stakeholder's role and limitations.

The Peruvian model of interagency workshops demonstrates how sustained collaboration can yield tangible results in the fight against illegal logging (Brazilian Institute of Environment, 2020). These workshops create structured platforms for police, forestry officials, and environmental NGOs to share intelligence, align enforcement priorities, and develop standardized operating procedures. What makes this approach particularly effective is its emphasis on after-action reviews, wherein participants analyze real cases to identify systemic weaknesses in the enforcement chain. This continuous learning process has led to measurable improvements in seizure rates and prosecution outcomes for illegal timber operations in the region.

Emerging best practices in collaborative training now emphasize the inclusion of non-traditional partners, such as financial investigators and corporate compliance officers. Environmental crimes increasingly intersect with complex financial crimes, requiring officers to understand how to follow money trails and identify shell companies used to launder the proceeds of illegal activities. Training programs in the European Union now routinely include modules on financial investigation techniques specifically tailored to environmental crimes, reflecting the evolving enforcement landscape (EU Environmental Enforcement Network, 2019).

8 Technological Integration for Next-Generation Enforcement

The digital transformation of law enforcement training has opened new frontiers in preparing officers to meet environmental crime challenges. Canada's innovative use of virtual reality for oil spill investigation training (Canadian Wildlife Federation, 2022) represents a paradigm shift in experiential learning. These VR modules allow officers to practice complex tasks, such as sampling contaminated sites or interviewing witnesses, in hyper-realistic simulations that would be impossible or dangerous to recreate physically. The technology's ability to generate multiple scenario variations ensures that officers develop adaptable skills for diverse field situations.

China's application of AI in predicting illegal waste dumping patterns (Chinese Ministry of Environmental Protection, 2023) illustrates how big data transforms environmental enforcement strategies. Training programs now incorporate data analytics modules, where officers learn to interpret predictive models and translate algorithmic outputs into targeted patrol strategies. This technological approach has proven particularly valuable in addressing the "whack-a-mole" nature of waste crime, where offenders frequently change their dumping locations to evade detection.

Blockchain technology is emerging as a powerful tool in the fight against environmental crime, particularly in verifying supply chain integrity. Training initiatives in several Southeast Asian countries now include blockchain certification programs for officers investigating the illegal wildlife trade (UNODC, 2021). These programs teach law enforcement personnel how to trace and verify transactions involving wildlife products, helping them distinguish between legal and illegal specimens in trade. The tamper-proof nature of the technology makes it particularly valuable for maintaining evidentiary chains in complex transnational cases.

9 Regional Case Studies: Contextualizing Global Challenges

The diversity of environmental crime manifestations across regions underscores the importance of localized training approaches. In Southeast Asia, training programs targeting illegal wildlife trafficking have evolved to address the region's unique role as both a source and transit hub for global trade (Interpol, 2021). These programs emphasize network analysis techniques to map trafficking routes and identify choke points for interdiction. A notable innovation has been the incorporation of behavioral psychology principles to help officers anticipate and disrupt trafficker adaptation strategies when traditional routes are compromised.

Brazil's satellite-based training for Amazon deforestation monitoring (IBAMA, 2020) demonstrates how technology can overcome the challenges of policing vast and remote areas. Officers receive intensive instruction in interpreting near-real-time satellite imagery, with training scenarios simulating the complete enforcement cycle from detection to ground verification. This approach has significantly reduced response times to illegal clearing activities while improving the quality of the evidence for prosecution. The program's success has inspired similar initiatives in other forest-rich nations facing deforestation.

Eastern Europe's experience with illegal coal mining presents a different set of enforcement challenges, particularly regarding the intersection of environmental crime and organized labor (EU Environmental Enforcement Network, 2019). Poland's training programs address these complexities through specialized modules on mining law, labor relations, and underground safety protocols. Diagnostic assessments have revealed that officers often lack technical knowledge about mining operations, leading to the development of hands-on training at decommissioned sites. This practical approach has improved the officers' ability to distinguish between legal artisanal mining and illegal industrial-scale operations.

These regional case studies collectively highlight the need for effective training programs to balance universal enforcement principles with localized adaptations. The most successful initiatives share common features: they are evidence-based, regularly updated to reflect changing crime patterns, and incorporate feedback loops from field officers to the training designers. As environmental crimes continue to evolve in sophistication, so too must the training approaches that prepare law enforcement agencies to combat them. The integration of collaborative frameworks with cutting-edge technologies represents the most promising path for building enforcement capacity on a global scale.

10 Challenges and Opportunities in Environmental Crime Enforcement Training

10.1 Persistent Challenges in Capacity Building

Resource constraints remain one of the most significant barriers to effective environmental law enforcement training globally. Many agencies in developing nations operate with severely limited budgets, often prioritizing traditional crime over environmental protection (UNEP, 2020). This financial strain manifests as outdated training materials, insufficient field equipment, and the inability to retain specialized instructors. These cascading effects create knowledge gaps that criminal networks readily exploit. Open-access resources, such as UNEP's enforcement manuals and INTERPOL's e-learning platforms, have become vital stopgap measures, but their translation and localization for different jurisdictions remain an ongoing challenge. Even in wealthier nations, the specialized nature of environmental crime investigation often means that training programs compete for funding against higher-profile law enforcement priorities.

The legal complexity surrounding environmental crime creates another substantial training hurdle. Officers must navigate a labyrinth of overlapping international conventions, national statutes, and local ordinances that frequently contain contradictory provisions (Elliott, 2016). This is particularly acute in transboundary cases involving waterways or migratory species, where jurisdictional ambiguities can paralyze enforcement. Training programs struggle to keep pace with rapidly evolving legal landscapes, as seen in recent amendments to the Basel Convention on plastic waste trade or the CITES listings for newly endangered species. The problem is compounded in federal systems, where environmental enforcement responsibilities are divided between multiple levels of government, requiring officers to master distinct legal frameworks operating simultaneously.

11 Emerging Opportunities for Enhanced Enforcement

Global networks have emerged as powerful force multipliers in addressing these challenges. Partnerships with organizations such as INTERPOL's Environmental Security Programme and UNEP's Law and Environment Assistance Platform have created unprecedented opportunities for knowledge exchange and coordinated action (Interpol, 2021). These networks facilitate not only information sharing but also joint training exercises that build interoperability between national enforcement agencies. The recent establishment of regional environmental crime task forces in Southeast Asia, Africa, and Latin America demonstrates how such collaborations can enhance detection capabilities while pooling scarce resources to combat wildlife crime. Crucially, these networks also provide channels for frontline officers to contribute their field experiences directly to training curriculum development, ensuring that programs remain grounded in operational realities.

Technological advancements are simultaneously revolutionizing environmental enforcement training and operations. Drone technology, once prohibitively expensive, is now incorporated into routine training programs for habitat monitoring and illegal activity detection (Kenya Wildlife Service, 2022). AI applications range from automated camera trap image analysis to predictive algorithms that identify high-risk areas for illegal mining and dumping (Chinese Ministry of Environmental Protection, 2023). Perhaps the most transformative has been the advent of mobile enforcement applications that put real-time legal databases, species identification guides, and evidence documentation protocols directly into officers' hands during field operations. These tools are increasingly integrated into training simulations, creating a seamless transition from the classroom to field applications.

12 Bridging the Implementation Gap

The path from recognizing these opportunities to realizing their full potential faces several challenges. Technological solutions often require substantial upfront investments in both equipment and digital literacy training, which many agencies cannot afford. There is also a risk of technological solutionism, where flashy tools are adopted without proper consideration of maintenance needs or their actual utility in local contexts. Successful programs, such as Brazil's satellite monitoring initiative, demonstrate the importance of pairing technology investments with comprehensive training on data interpretation and chain-of-custody procedures (IBAMA, 2020).

Similarly, although global networks provide invaluable support, they cannot replace the need for localized training approaches that account for specific ecological, cultural, and legal contexts. The most effective programs strike a careful balance between adopting international best practices and adapting them to local conditions. Vietnam's hybrid approach to wildlife crime training, which combines INTERPOL's standardized investigative techniques with region-specific knowledge about local trafficking networks, exemplifies this balanced model (UNODC, 2021).

As environmental crimes grow in sophistication and scale, training programs preparing officers to combat them must evolve accordingly. The challenges are substantial, but the expanding toolkit of collaborative frameworks and technological solutions offers unprecedented opportunities for building enforcement capacity. The coming decade will likely see an increased emphasis on climate-related crimes and digital environmental offenses, requiring continuous adaptation of training curricula to address these emerging threats. What remains constant is the fundamental need for well-trained and properly equipped officers to serve as the frontline defense against environmental degradation.

13 Conclusion

Environmental crime training for law enforcement officers must fundamentally reorient itself to prioritize both ecological protection and human dignity. The complex nature of these offenses demands more than technical proficiency; it requires a profound understanding of the interconnected web of social, economic, and environmental consequences that these crimes create. As this chapter has demonstrated, effective training programs must serve dual purposes: equipping officers with rigorous investigative skills and fostering deep empathy for affected communities and ecosystems. This humanistic approach transforms environmental enforcement from a purely regulatory function into a form of social service that protects both people and the planet.

The pedagogical framework outlined in this chapter represents a significant evolution beyond the traditional law enforcement training models. By integrating criminological theory with hands-on field experience, officers develop not only the ability to respond to environmental crimes but also the capacity to anticipate and prevent them. The interdisciplinary nature of this training, spanning legal, scientific, economic, and ethical domains, creates enforcement professionals capable of addressing environmental offenses in their full complexity. This comprehensive approach is particularly crucial as environmental crimes increasingly intersect with other serious offenses, including organized crime, corruption, and human rights violations.

Regional case studies from across the globe reveal both universal principles and localized adaptations necessary for effective training. While the illegal wildlife trader in Southeast Asia, Amazon logger, and Eastern European coal miner may operate in vastly different contexts, they share common motivations and methods that training programs must address. The most successful initiatives combine standardized investigative techniques with deep cultural and ecological knowledge specific to their operating environment. This balance between global best practices and local implementation creates enforcement strategies that are both principled and practical in nature.

Technological innovation presents unprecedented opportunities to enhance the effectiveness and scalability of training. From virtual reality simulations that recreate complex crime scenes to AI-powered predictive policing tools, these advancements allow officers to train more realistically and frequently than ever. However, as this chapter cautions, technology must serve as an enabler rather than a replacement for fundamental enforcement skills and human judgment. The most impactful training programs use technology to amplify, not diminish, officers' connection to the communities and environments they protect.

Looking ahead, environmental law enforcement training must remain agile in the face of emerging challenges. Climate change creates new categories of environmental offenses while intensifying existing ones. Digital technologies enable novel forms of environmental crime through cryptocurrency-enabled trafficking and cyber-enabled pollution concealment. Training curricula must evolve continuously to address these developments while maintaining a focus on the core principles of justice, sustainability, and human rights.

Ultimately, the measure of success for these training programs lies not just in arrest statistics or seizure numbers, but in their ability to create lasting environmental protection and social benefits. Well-trained officers become force multipliers, educating communities, deterring potential offenders, and building institutional capacity that extends far beyond individual cases. By investing in comprehensive, compassionate, and cutting-edge training for environmental law enforcement, societies make a profound statement about the value they place on ecological integrity and social justice. The challenges are significant, but the opportunity to shape a new generation of environmental stewards within law enforcement represents one of the most promising strategies for creating a more sustainable and just world.

References

Australian Environmental Protection Authority. (2022). *River pollution detection training modules*. Commonwealth of Australia.

Brazilian Institute of Environment and Renewable Natural Resources. (2020). *Amazon mining enforcement report*. IBAMA Press.

Canadian Wildlife Federation. (2022). *Virtual reality applications for wildlife crime investigations: Technical manual*. CWF Publications.

Chinese Ministry of Ecology and Environment. (2023). *Artificial intelligence applications in environmental enforcement: Implementation guidelines*. MEE Press.

Clarke, R. V. (1997). *Situational crime prevention: Successful case studies* (2nd ed.). Harrow and Heston.

Elliott, L. (2016). *The global politics of the environment: Environmental crime and justice* (3rd ed.). Routledge.
European Union Environmental Enforcement Network. (2019). *Case studies in transboundary environmental enforcement.* EU Publications Office.
Interpol. (2021). *Guide to environmental crime investigations.* INTERPOL Environmental Security Programme.
Kenya Wildlife Service. (2022). *GIS applications for wildlife enforcement: Training handbook.* KWS Publications.
Lynch, M. J., Stretesky, P. B., & Long, M. A. (2018). *Green criminology: Crime, justice, and the environment*. University of California Press.
Philippine National Police Maritime Group. (2021). *Coastal environmental crime training programs: Operational manual.* PNP Publications.
Slovenia Ministry of Environment and Spatial Planning. (2022). *Water pollution investigation manual.* MOESP Press.
United Nations Environment Programme. (2020). *Environmental crime report: Emerging challenges and innovative responses.* UNEP.
United Nations Office on Drugs and Crime. (2021). *Wildlife and forest crime analytical toolkit* (2nd ed.). UNODC.
White, R. (2014). *Environmental crime and justice: A reader*. Palgrave Macmillan.

Dr. Mamdooh Abdelhameed Abdelmottlep is a former law enforcement officer with extensive professional experience in police training, security education, and international academic cooperation. In addition to founding the International Police Science Association (IPSA), Dr. Mamdooh created the World Internal Security and Police Index (WISPI) and Global Police and Security Index (GPSI), which measure the capacity and efficiency of police and security service providers to address internal security issues. Dr. Mamdooh's work has earned him numerous professional honors, including the 2014 CRIMINOLOGIA.IT, awarded by the United Nations, and the 2016 Distinction Medal, awarded by Sharjah's Ruler, His Highness Sheikh Dr. Sultan Al-Qasimi, for the continuous development of Middle Eastern Police Science and International Security Education. His research focuses on police science, law enforcement, and criminal justice administration.

Chapter 6
Environmental Protection During Armed Conflicts Between Legal Codification and Judicial Practice

Adil El Balbal and Abdelaali Arroubi

1 Introduction

Armed conflict leaves more than human casualties in its wake; it scars the land itself. Field assessments by the United Nations Environment Programme have traced oil-slicked rivers, poisoned soils, and razed forests to recent wars, showing how violence can unravel entire ecosystems (UNEP, 2009). Yet when hostilities break out, commanders concentrate on territorial gains and political leverage, not on wetlands or woodlands caught in the cross-fire. This blindness has spurred diplomats and lawyers to widen the rules of war. The Hague Regulations of 1899 and 1907 first hinted that belligerents owe duties beyond the battlefield. The Additional Protocols to the Geneva Conventions, adopted in 1977, went further, prohibiting acts that cause "*widespread, long-term and severe*" damage to the natural environment. Subsequent environmental treaties have reinforced that message, signaling growing consensus that military necessity is not a license to ravage the planet.

Today, environmental protection sits squarely within international humanitarian law. The challenge is to turn these clauses into concrete practice—so when the guns finally fall silent, rivers can run clear, soils can sustain crops, and ecosystems can begin to heal.

Legal language now pays far more attention to nature than it once did. Environmental provisions have filtered into the laws of war, and a handful of international courts and tribunals have begun testing the idea that those provisions carry real consequences. The International Court of Justice, ad-hoc arbitral panels, and even a few

A. E. Balbal (✉) · A. Arroubi
Faculty of Law and Economics, Moulay Ismail University, Meknes, Morocco
e-mail: a.elbalbal@edu.umi.ac.ma

A. Arroubi
e-mail: Abdelaaliarr@gmail.com

M. Mliless et al. (eds.), *Frontiers of Environmental Law Enforcement*,
https://doi.org/10.1007/978-3-032-10205-8_6

criminal tribunals have all entertained claims that link armed conflict to ecological damage. Still, theory has run ahead of practice.

Take Article 55 of Additional Protocol I (ICRC, 1977), it obliges parties to protect the natural environment from warfare that risks "widespread, long-term or severe" harm. On paper, the clause looks robust; on the ground, it has rarely been invoked, let alone enforced. A parallel gap appears on the environmental-law side. Cornerstone treaties such as the Convention on Biological Diversity or the Basel Convention were drafted for peacetime and say almost nothing about what happens when fighting breaks out. As Crawford observes in *Brownlie's Principles of Public International Law*, this split leaves a "legal vacuum" whenever war and ecology collide (Crawford, 2019).

Criminal accountability is equally thin. The Rome Statute technically allows the International Criminal Court to pursue environmental devastation, yet prosecutions remain exceptional. Broader forums of international justice have ruled on pollution or resource disputes, but the case law is sparse and enforcement tools are weak. Without credible sanctions, the deterrent effect is modest.

Against that backdrop, this study asks how well the existing patchwork actually works. Do the rules of humanitarian law, together with mainstream environmental treaties, provide real shields for ecosystems in wartime—or do loopholes and jurisdictional blind spots cancel out their promise? By reading the key conventions alongside the most relevant court and arbitral decisions, the research maps successes, shortcomings, and structural hurdles. Within this framework, the central research question guiding this study is:

> To what extent do current international norms and judicial precedents protect the environment during armed conflict, and what obstacles prevent those protections from taking full effect?

This investigation begins by asking where the international community truly stands. It reviews the existing web of rules that address environmental harm in wartime, then considers how often international prosecutors have pursued ecological destruction and why those efforts so frequently lose momentum. It also explores whether broader tribunals—those that hear civil or inter-state claims—actually provide meaningful remedies for wartime environmental damage. From there, the discussion confronts the legal loopholes, evidentiary obstacles, and political pressures that blunt the force of today's protections and, finally, considers what concrete steps might empower courts and arbitral panels to shield nature even when armed conflict erupts.

To tackle these issues, the study relies on a layered approach. Close textual analysis of treaties, protocols, and judicial opinions establishes the doctrinal baseline; a comparative lens then sets the different accountability models—criminal, civil, and inter-state—side by side to assess their practical reach. Throughout, a critical perspective tests elegant legal principles against the messy realities recorded in case files and field reports.

The argument unfolds in two parts. The first maps the legal scaffolding: the provisions of humanitarian law, the parallel corpus of environmental treaties, and the points at which they intersect or clash. The second moves from law to courtroom practice, gauging how international judges and arbitrators have applied—or sidestepped—those rules in actual disputes. The study concludes by surveying the practical and political barriers that still leave ecosystems exposed and by outlining reforms that might transform wartime environmental protection from an aspirational ideal into an enforceable norm.

2 A Humanitarian Law as a Safety Net for Nature

Decades of research—from the World Charter for Nature (United Nations General Assembly. (1982) and Kiss's (1989) commentary on it, to recent analyses by Bouvier (1991) and Howell (2023)—have kept one fact in sharp focus: war maims the natural world in ways that can outlast any peace treaty. Forests burn, aquifers turn toxic, and whole ecosystems slide past the point of recovery, jeopardizing both current populations and generations yet to come.

These hard lessons have pushed diplomats and jurists to weave environmental concerns into the fabric of international humanitarian law (IHL). Taken together, the Hague regulations on the conduct of hostilities and the Geneva provisions on the protection of victims form a body of conventional and customary norms that now address ecological harm as well as human suffering. Faure (2022) notes that IHL accomplishes this in two ways. Some rules speak directly to environmental damage—prohibiting methods of warfare likely to cause widespread, long-term or severe harm—while others offer indirect safeguards by limiting the weapons, tactics or targets that belligerents may lawfully choose.

No one pretends these constraints can eradicate every act of wartime pollution or habitat loss. Instead, the goal is to force military planners to keep destruction within what the law still reluctantly calls "*acceptable*" bounds. The system is imperfect, but it represents a collective recognition that a healthy environment is not a collateral luxury; it is, as Kiss argued back in 1988, a third-generation human right integral to the survival and dignity of all people.

2.1 Indirect Protection of Nature in Humanitarian Law

International humanitarian law rarely lists "the environment" as a distinct category, yet many of its rules, both customary and conventional, end up shielding ecosystems by imposing limits on how wars are fought. These restraints act not through a single, self-contained regime but through a network of principles that restrict methods and means of warfare likely to inflict serious ecological harm.

Customary Rules for Environmental Protection During Armed Conflicts

Customary IHL sets out several interlocking principles that, together, form a protective web around the natural environment. The principle of limitation is the first strand. From the St Petersburg Declaration (St Petersburg Declaration). (1868b) to Article 35(1) of Additional Protocol I (1977), belligerents are reminded that they may not choose weapons or tactics at will; some options are off-limits because their effects are inherently indiscriminate or cruel (Solis, 2010).

Running parallel is the principle of proportionality. Long accepted as customary law and codified in Rule 43 of the ICRC study, this rule forbids attacks where the expected ecological damage would be excessive in relation to the concrete and direct military advantage (Bothe, 1996; Henckaerts & Doswald-Beck, 2005).

The principle of distinction, also captured in Rule 43, reinforces that obligation. Combatants must differentiate military objectives from civilian objects—and from natural features that serve no military function—thereby outlawing the deliberate destruction of forests, rivers or farmland that are not legitimate targets.

Environmental immunity, grounded in the notion of military necessity, adds another layer of restraint. Unless environmental destruction is imperatively demanded by the necessities of war, it is prohibited; Article 35(3) of Additional Protocol I explicitly bars invoking necessity to justify acts likely to cause widespread, long-term, or severe damage (Simonds, 1992).

Rule 44 of the ICRC compilation obliges states to weigh environmental consequences when selecting weapons and tactics, while Rule 45 goes further, banning any method expected to produce widespread, long-term, and severe harm or to use environmental destruction as a weapon (Henckaerts & Doswald-Beck, 2005; Michael & Schmitt, 1997a, 1997b).

Beyond these specific rules, other customary norms originally designed to protect civilians—such as limits on bombardment or safeguards for food and water supplies—often spare ecosystems as well. Multilateral treaties that curb particularly destructive weapons, even when not framed as "green" accords, likewise reinforce the emerging consensus: nature must not become another casualty of war.

Agreements that Offer the Environment Indirect Protection in Wartime

Over time, states have negotiated a series of treaties that—although drafted for broader humanitarian reasons—end up shielding ecosystems by curbing how certain weapons may be used. Their common thread is a determination to outlaw methods of war whose knock-on effects on nature would be intolerable.

The story begins with the St Petersburg Declaration of 1868. By banning explosive projectiles under 400 grams, the drafters sought to spare combatants "unnecessary suffering," but the rule also spared surrounding landscapes from indiscriminate blast and fire (St Petersburg Declaration, 1868a). A few decades later, the Hague Conventions of 1899 and 1907 embedded the principles of distinction and proportionality—cornerstones that would later frame environmental debates. Article 23 of the 1907 Hague Regulations, for example, outlaws' weapons that cause needless injury, a prohibition that can be read today as shielding natural systems from gratuitous devastation (Henckaerts & Doswald-Beck, 2005).

The inter-war period pushed the envelope further. The 1925 Geneva Protocol condemned the battlefield use of asphyxiating, poisonous gases and of bacteriological agents, recognizing that toxins released against troops would inevitably poison soils, rivers, and wildlife (Arrassen, 1986). Fast-forward to 1993 and the Chemical Weapons Convention takes that logic to its conclusion: no development, no production, no stockpiling, and mandatory destruction—an acknowledgment of the "severe and lasting" environmental fallout that chemical arsenals inflict.

Similar concerns lie behind the campaign to restrain nuclear arms. The 1968 Nuclear Non-Proliferation Treaty and, more recently, the 2017 Treaty on the Prohibition of Nuclear Weapons show how far consensus has moved (UN, 2017): weapons capable of irradiating whole regions and contaminating ecosystems for centuries are seen as incompatible with any acceptable vision of security.

Traditional humanitarian law may not have set out to be an environmental charter, yet its cumulative principles—limiting suffering, forbidding indiscriminate harm—have evolved into an indirect but significant ecological shield. Still, as modern conflicts grow fiercer and environmental stakes rise, the need for explicit, purpose-built rules to protect nature in war has become ever more pressing.

2.2 Direct Protection of the Natural Environment During Armed Conflict

Detailed reading of modern humanitarian instruments shows that today's law moves beyond indirect safeguards and speaks directly to the need to spare ecosystems from the violence of war. Four pillars illustrate this trend: the 1976 Convention on the Prohibition of Military or Any Other Hostile Use of Environmental Modification Techniques (ENMOD), the 1977 Additional Protocol I to the Geneva Conventions, the ICRC's Guidelines on the Protection of the Natural Environment in Armed Conflict (ICCR, 1994, 2020), and the draft principles on the same theme adopted by the International Law Commission in 2022.

Treaty Rules that Place Explicit Limits on Wartime Harm to Nature

Concern about deliberate environmental manipulation grew sharply in the early 1970s, culminating in ENMOD, opened for signature on 10 December 1976 and in force since 5 October 1978. The treaty bans any military use of "environmental modification techniques"—from cloud-seeding designed to flood croplands to the triggering of earthquakes or tsunamis—recognizing that such tactics threaten planetary systems themselves (Bothe et al., 2013).

A year later, Additional Protocol I introduced parallel restraints. Article 35(3) forbids methods or means of warfare that are intended, or may be expected, to inflict "widespread, long-term and severe" damage on the natural environment, while Article 55 obliges belligerents to protect the environment against precisely that scale of harm (ICRC, 1977). During the drafting of Additional Protocol II, delegates even debated a proposed Article 28 bis that would have extended similar language to non-international conflicts. Although the amendment failed at the final plenary, the debate itself signaled a growing conviction that environmental stewardship should survive the onset of hostilities.

Outside the treaty arena, the ICRC and the International Law Commission have carried the project further. The ICRC's, 2020 Guidelines distil existing obligations and good practice, offering militaries concrete instructions for integrating environmental concerns into operational planning. Meanwhile, the ILC's (2022) draft principles marshal both humanitarian and environmental law to spell out a coherent set of duties, from prevention to post-conflict restoration (UNEP & ILC, 2022). Together, these texts underscore an emerging consensus: even in war, nature is not a legitimate casualty.

Recent Initiatives that Consolidate Direct Protection

Momentum has gathered only in the past decade, yet the field has already registered two landmark achievements. The first milestone is the work of the International Law Commission. After more than a decade of study, the Commission finalized its Draft Principles on the Protection of the Environment in Relation to Armed Conflicts, which the General Assembly "welcomed" in resolution 77/104 of 7 December 2022. The text sets out 27 principles that knit together duties owed before, during, and after hostilities. Principle 2 states the overarching purpose, Principle 12 reaffirms a broad obligation to spare the environment while fighting, and later provisions allocate responsibility for remedial measures once the guns fall silent (UNGA, 2022).

A second pillar appeared when the ICRC issued its Guidelines on the Protection of the Natural Environment in Armed Conflict (ICCR, 2020), an expanded successor to the organization's 1994 recommendations. Drawing on treaty and customary humanitarian law, the document gathers 32 rules and good-practice notes. Part 1 singles out categories of special protection, Part 2 sets out general safeguards, and Part 3 addresses particular weapon systems. Rule 28, for example, stresses states' duty to prosecute war crimes that inflict serious environmental damage (ICRC, 2020).

These parallel efforts—from the ILC's codification project to the ICRC's operational guidance—cement the idea that shielding ecosystems is a non-negotiable component of the modern laws of war. Yet normative advance naturally leads to the next question: do existing judicial forums possess the reach and the resolve to deter, adjudicate, and remedy grave environmental harm committed in wartime?

3 International Judicial Protection of the Environment During Armed Conflict

In contemporary practice, "international judicial protection of the environment" describes an emerging effort to criminalize and punish grave ecological harm—especially harm that spills across borders, is linked to armed violence, or threatens the collective interests of humanity. For decades, public international law contained no express crime of environmental destruction, but successive treaty negotiations, tribunal statutes, and judicial opinions have begun to close that gap. Today, international justice—whether delivered through criminal courts, arbitral panels, or hybrid mechanisms—constitutes a core avenue for accountability when ecosystems are damaged in war. International judicial protection of the environment refers to the global efforts to criminalize and punish acts that cause serious ecological harm, particularly those committed during armed conflicts, that cross-national borders, or that threaten international peace and security and the collective interests of humanity. While public international law did not originally include explicit provisions criminalizing environmental damage, developments over recent decades have led to significant attempts to close this legal gap. These include incorporating environmental protection into the mechanisms of international criminal law. In this context, international justice has emerged as a vital tool to ensure accountability for environmental destruction linked to armed conflict—whether through international criminal courts, arbitration mechanisms, specialized tribunals, or general judicial bodies such as the International Court of Justice (ICJ). In its 1996 *Legality of the Threat or Use of nuclear weapons* advisory opinion, the ICJ underscored that states must take environmental considerations into account when assessing the lawfulness of particular means of warfare, signaling that environmental impacts are relevant to the interpretation of humanitarian-law principles (ICJ, 1996).

3.1 Ad-Hoc Criminal Tribunals and the First Steps Toward Accountability

The need to treat ecological devastation as more than a collateral issue has become apparent as conflicts, industrial disasters, and climate emergencies multiply. Earlier international tribunals did not create stand-alone environmental offences, yet their jurisprudence hints at an implicit concern for wartime ecology.

The Nuremberg proceedings offer a useful starting point. Although environmental harm never appeared as a separate count, several trials touched on conduct that modern law would recognize as ecologically criminal. In *United States v. Carl Krauch et al.* (the IG Farben case), the defendants were prosecuted for supplying Zyklon-B and for constructing vast chemical plants—most infamously Monowitz near Auschwitz—that polluted the surrounding land and water while relying on confiscated property (Trials of War Criminals before the Nuremberg Military Tribunals,

1952, Vol. VIII). Likewise, *United States v. Friedrich Flick et al.* dealt with illegal exploitation of mines and raw materials; today, such large-scale resource pillage would likely be framed as an environmental crime (Nuremberg Military Tribunals, 1952).

Half a century later the Security Council created the International Criminal Tribunal for the former Yugoslavia (ICTY). Its mandate did not single out environmental offences, but several cases nonetheless exposed the ecological cost of military operations. The indictment against Slobodan Milošević referenced the bombing of oil refineries, power stations, and chemical plants whose combustion released toxins into air, soil, and waterways (ICTY, *Prosecutor v. Milošević*, IT-02-54, 2001). In *Prosecutor v. Kordić & Čerkez*, the judges recorded the destruction of forests, reservoirs and water-supply networks—damage folded into broader charges of unlawful attacks on civilian objects (ICTY, 2001).

Taken together, Nuremberg and the ICTY did not create an autonomous category of "environmental war crime," but they cemented the principle that the natural world cannot be an unregulated target. Their reasoning paved the way for the International Criminal Court (ICC) to contemplate explicit ecological liability.

3.2 The Rome Statute and the ICC's Latent Capacity

The adoption of the Rome Statute in 1998 signaled a decisive shift. Article 8(2)(b)(iv) categorizes as a war crime any attack that causes "widespread, long-term and severe" environmental damage when the harm is clearly excessive to the anticipated military advantage (Rome Statute, 1998). The threshold is demanding, but the clause establishes a direct juridical basis for future prosecutions. Environmental devastation can also appear under crimes against humanity when linked to forcible displacement, starvation tactics, or the destruction of life-sustaining resources.

Practical application, however, has been limited. No ICC case has centered exclusively on environmental harm, although the Court has acknowledged such damage obliquely. In *The Prosecutor v. Jean-Pierre Bemba Gombo*, the Office of the Prosecutor mentioned the ecological fallout from assaults on civilian infrastructure in the Central African Republic (ICC, n.d). Former Prosecutor Fatou Bensouda, in her policy papers, repeatedly flagged environmental destruction—especially resource exploitation that fuels conflict—as a matter the Court might pursue (ICC, 2016).

Unlocking the Rome Statute's full potential will depend on several conditions. Judicial interpretation of war-crime and crimes-against-humanity provisions must be willing to foreground ecological evidence. States Parties must cooperate in gathering the technical data needed to prove complex environmental harm. Above all, governments must show the political nerve to support prosecutions even when powerful actors are implicated.

4 Conclusion

The preceding analysis shows that legal and judicial safeguards for nature in war remain fragile. International rules—whether Additional Protocol I, the ENMOD Convention, or Article 8 of the Rome Statute—contain promising language, yet implementation is hampered by ambiguous terms such as "widespread," "long-term," and "severe," by difficult evidentiary demands, and by the absence of binding enforcement machinery. Environmental devastation still tends to be treated as an incidental effect of other crimes rather than a distinct wrong. Meaningful progress will require a concerted overhaul. Treaty provisions should be clarified and expanded: Article 8 could be amended so that the three-part test is not an all-or-nothing barrier, and deliberate, large-scale ecological destruction—sometimes described as ecocide—should be codified as a separate international crime. An independent monitoring body under UNEP could gather forensic data, while environmental expertise ought to be embedded in UN investigative teams and peace operations. Military training and rules of engagement must internalize a "do-no-environmental-harm" standard. Finally, the ICC itself could prioritize cases where environmental damage is central, demonstrating that devastation of ecosystems is not an acceptable cost of war. Only when these legal, institutional, and political steps converge will international humanitarian law deliver the level of environmental protection that present circumstances—and future generations—demand.

References

Arrassen, M. (1986). *Conduite des hostilités: droit des conflits armés et désarmement* [Conduct of hostilities: law of armed conflict and disarmament]. Bruylant.

Bothe, M. (1996). War and environment. In R. Bernhardt (Ed.), *Encyclopedia of Public International Law, 4*, 291–295.

Bothe, M., Partsch, K. J., & Solf, W. A. (2013). *New rules for victims of armed conflicts: Commentary on the two 1977 Protocols additional to the Geneva conventions*. Martinus Nijhoff.

Bouvier, A. (1991). La protection de l'environnement naturel en période de conflit armé [Protection of the natural environment in times of armed conflict]. *Revue Internationale De La Croix-Rouge, 792*, 599–611.

Crawford, J. (2019). *Brownlie's principles of public international law* (9th ed.). Oxford University Press. https://doi.org/10.1093/he/9780198737445.001.0001

Faure, C. (2022). *Manuel de droit des opérations militaires* [Manual of Military Operations Law]. Direction des affaires juridiques, État-major des armées. Retrieved from: https://www.defense.gouv.fr/sites/default/files/sga/Manuel%20de%20droit%20des%20op%C3%A9rations%20militaires_%C3%A9dition%202022.pdf

Henckaerts, J. M., & Doswald-Beck, L. (2005). *Customary international humanitarian law* (Vol. I). Cambridge University Press.

Howell, J. (2023). Impact environnemental des conflits armés [Environmental impact of armed conflicts]. *Commission des questions sociales, de la santé et du développement durable*, Doc. 15674. Retrieved from: https://assembly.coe.int/nw/xml/XRef/Xref-XML2HTML-fr.asp?fileid=31460&lang=fr Accessed on 5 January 2023

ICC Office of the Prosecutor. (2016). *Policy paper on case selection and prioritisation.* Retrieved from: https://www.icc-cpi.int/sites/default/files/itemsDocuments/20160915_OTP-Policy_Case-Selection_Eng.pdf
ICC. (n.d.). *The Prosecutor v. Jean Bemba Gombo* (ICC-01/05–01/08), Case information sheet. Retrieved from: https://www.icc-cpi.int/sites/default/files/CaseInformationSheets/BembaEng.pdf
ICJ. (1996). *Legality of the threat or use of nuclear weapons*, Advisory opinion. Retrieved from: https://iilj.org/wp-content/uploads/2016/08/Legality-of-the-Threat-or-Use-of-Nuclear-Weapons-1996.pdf
ICJ. (1997). *Case concerning the Gabcikovo-Nagymaros Project (Hungary/Slovakia)*, Judgment of 25 September. Retrieved from: https://www.icj-cij.org/case/92
ICRC. (1977). Protocol Additional to the Geneva Conventions of 12 August 1949, and relating to the protection of victims of international armed conflicts (Protocol I), (Art. 55). Retrieved from: https://ihl-databases.icrc.org/en/ihl-treaties/api-1977) Accessed on 8 June 1977
ICRC. (1994). *Guidelines on the protection of the environment*, 25–30. Retrieved from: https://www.icrc.org/en/doc/resources/documents/resolution/council-delegates-resolution-1-1994.htm
ICRC. (2020). *Guidelines on the protection of the natural environment in armed conflict*. Retrieved from: https://www.icrc.org/sites/default/files/document_new/file_list/guidelines_on_the_protection_of_the_natural_environment_in_armed_conflict_advance-copy.pdf Accessed on July 25, 2025
ICTY (2001). *Prosecutor v. Dario Kordić & Mario Čerkez* (Judgment, Case No. IT-95–14/2-T). Retrieved from: https://www.icty.org/x/cases/kordic_cerkez/tjug/en/kor-tj010226e.pdf Accessed on 26 February 2001
ILC. (2022). *Draft principles on protection of the environment in relation to armed conflicts* (Principles 13–15). Retrieved from: https://legal.un.org/ilc/reports/2022/
Kiss, A. (1989). *Droit international de l'environnement. Revue Juridique de l'Environnement, 4.* Retrieved from: www.persee.fr/doc/rjenv_0397-0299_1989_num_14_4_2558_t1_0479_0000_1
Schmitt, M. N. (1997a). Green war: An assessment of the environmental law of international armed conflict. *Yale Journal of International Law, 22*, 1–52.
Nuremberg Military Tribunals. (1952). *Trials of War Criminals before the Nuremberg Military Tribunals under Control Council Law No. 10: The Flick Case* (Vol. VI). Retrieved from: https://tile.loc.gov/storage-services/service/ll/llmlp/2011525364_NT_war-criminals_Vol-VI/2011525364_NT_war-criminals_Vol-VI.pdf
Rome Statute of the International Criminal Court. (1998, July 17). *United Nations Treaty Collection.* Retrieved from: https://treaties.un.org/doc/Publication/UNTS/Volume%202218/volume-2218-I-38544-English.pdf
Schmitt, M. N. (1997b). Green war: An assessment of the environmental law of international armed conflict. *Yale Journal of International Law, 22*(1), 1–52.
Simonds, S. N. (1992). Conventional warfare and environmental protection: A proposal for international legal reform. *Stanford Journal of International Law, 29*(1), 123–160.
Solis, G. D. (2010). *The law of armed conflict: International humanitarian law in war*. Cambridge University Press.
St. Petersburg Declaration. (1868a). *Declaration banning the military use of certain explosive projectiles*. https://ihl-databases.icrc.org/en/ihl-treaties/st-petersburg-decl-1868 accessed on 29 November 1868
St Petersburg Declaration). (1868b, December 11). *International Committee of the Red Cross—Treaties, States Parties and Commentaries.* Retrieved from: https://ihl-databases.icrc.org/en/ihl-treaties/decl-stpetersburg-1868
UN. (2017). Treaty on the prohibition of nuclear weapons. Retrieved from: https://treaties.un.org/Pages/ViewDetails.aspx?src=TREATY&mtdsg_no=XXVI-9&chapter=26&clang=_en
UNEP & ILC. (2022). *Report on the work of its seventy-third session (U.N. Doc. A/77/10). United Nations.* Retrieved from: https://undocs.org/A/77/10

UNEP. (2009). *Protecting the environment during armed conflict: An inventory and analysis of international law*. Retrieved from https://www.unep.org/resources/report/protecting-environment-during-armed-conflict-inventory-and-analysis-international

United Nations General Assembly. (1982). World Charter for Nature (A/RES/37/7). Retrieved from: https://digitallibrary.un.org/record/39295?ln=en

United Nations General Assembly. (2022). *Protection of the environment in relation to armed conflicts* (Resolution 77/104, 7 December 2022). United Nations. Retrieved from: https://undocs.org/en/A/RES/77/104

Adil El Balbal is a PhD candidate at the Laboratory of Legal, Political, and International Studies at the Faculty of Legal, Economic, and Social Sciences, Moulay Ismail University, Meknes, Morocco. He holds a Master's degree in Security Governance and Human Rights from the same institution. His academic interests focus on international humanitarian law, with particular emphasis on the environmental impacts of armed conflicts and the broader issues of environmental security.

Abdelaali Arroubi is a PhD candidate at the Laboratory of Legal, Political, and International Studies (LEJPI), within the Faculty of Legal, Economic, and Social Sciences at Moulay Ismail University in Meknes, Morocco. He holds a Master's degree in Public Policy and conducts research at the intersection of administrative and environmental law. His work explores the relationship between administrative law and human rights, particularly in the context of migration governance, focusing on the rights and obligations of migrants and the legal interactions between transit and destination countries. He also investigates legal frameworks for biodiversity conservation, with particular attention to forest and wildlife protection.

Chapter 7
Green International Law Enforcement: Challenges and Opportunities

Rabiaa Bourhim

1 Introduction

International environmental law, also known as 'green international law', not only presents numerous challenges but also offers significant opportunities to protect the global environment. This constantly evolving field addresses issues such as climate change, biodiversity loss, and transboundary pollution. Its objective is to promote sustainable development, preserve natural resources, and prevent environmental degradation. It should be settled from the start that Green Environmental Law is defined in Encyclopedia Britannica (Shaw, 2025) that Green International Law, or International Environmental Law, refers to the body of international rules, principles, and agreements designed to manage and protect the global environment by addressing issues like climate change, ozone depletion, and biodiversity loss. It sets collective responsibilities and facilitates cooperation among states to achieve global environmental goals, regulate harmful activities such as pollution and hazardous waste trafficking, and provide mechanisms like the Green Climate Fund for climate justice and adaptation in vulnerable countries.

One of the main problems with international law in the context of international relations is that it conflicts with the principles of state sovereignty and non-interference in their affairs. This reduces its binding nature, thereby limiting its application and often rendering it insufficient. This is why the evolution, implementation, and enforcement of international law remain subject to the agreement of the states involved.

In our study, we will examine the role of international environmental law as a branch of international law in the application and enforcement of law on states. This is particularly relevant given the leading role now played by international non-governmental organisations and individuals. We will also consider how this influence

R. Bourhim (✉)
Doctor of Public Law and Political Sciences, Casablanca, Morocco
e-mail: Rabie.bourhim@gmail.com

M. Mliless et al. (eds.), *Frontiers of Environmental Law Enforcement*,
https://doi.org/10.1007/978-3-032-10205-8_7

has become effective in the application of international environmental law, and consequently international law in general. This study also aims to analyse the relationship between soft and hard law, and why states might adopt soft law more frequently to regulate environmental issues. International judicial and arbitration mechanisms for dispute resolution have significantly contributed to the development of international environmental law, particularly with regard to general concepts such as sustainable development and the principles of cooperation, prevention, the polluter pays principle, and the precautionary principle, as well as transboundary pollution (Tullio, 2010). Environmental standards are characterised by blurred legal boundaries that transcend the traditional framework of source theory. Consequently, studying them offers a new perspective on the concept of international normativity. Environmental standards are also based on legal uncertainty, which complicates the determination of their legal status and how they interact with other legal rules. International environmental law is a particularly interesting subject within the field of practical international law analysis. Put simply, international environmental law is international law applicable to environmental issues.

In reality, there are few legal precedents that explicitly rely on international environmental law. Nevertheless, international law is playing an increasingly important role in how judges approach environmental issues.

2 Changing Role of NGOs in Environmental Law

According to environmental experts, international environmental law is characterised by two failures: the ineffectiveness of the standard-setting process due to slow or even paralyzed negotiations, and difficulties at the implementation stage due to a lack of effective monitoring and enforcement mechanisms. In practice, however, civil society plays an important role in environmental negotiations. At the 1992 Rio Conference, for example, more than 20,000 NGO representatives were present. At COP21, specific spaces were reserved for non-state actors, including: NGOs, businesses, and local authorities.

However, the law lags behind this reality. While the principle of public participation is indeed enshrined in certain international texts, such as the Aarhus Convention, it is, paradoxically, only for the development of internal standards. Furthermore, civil society participation varies across international forums. The requirement for participatory democracy must be transposed to the international level. The Environment Committee believes that institutionalising the role of civil society would sustain and strengthen this type of initiative by giving it greater legitimacy. It therefore proposes the following:

– *implement a citizens' initiative and a universal right of petition to international institutions; and*

– *adopt an international convention on public participation in the development of international environmental law, based on the Almaty guidelines. This could clarify and consolidate the fundamental principles in this area, including public participation, access to information held by international organisations, and accreditation rules* (Aguila, 2015).

NGOs have experienced extraordinary growth in the field of environmental law, contributing to its rapid evolution in recent years. The emergence of NGOs has indeed led to collective issues meaningful to society as a whole being addressed. Thanks to the work of NGOs, citizens can learn about current issues in various fields and influence the priorities set by their authorities if they wish:

> Given the importance of NGOs as pressure groups, we can say that, through their direct, non-violent and courageous actions and by sacrificing their own integrity, they have achieved results of paramount importance. For example, following Greenpeace's actions, the United States stopped its nuclear testing campaign in Amchitka in 1971. In 1985, nuclear testing once again became the subject of international controversy when the French secret service sabotaged the Rainbow Warrior, Greenpeace's flagship. The latter was attempting to peacefully stop French nuclear testing (Fuentes Véliz, 2007, p. 412).

In the same context, Fuentes Véliz (2007) argues that NGOs function as pressure groups because they continuously draw public attention to environmental issues. Their role is to inform and raise awareness within civil society, which constitutes a crucial principle of their mission.

Another form of action carried out by NGOs as pressure groups is lobbying, i.e. trying to influence legislators or other public officials who have decision-making power. For instance, after many years of lobbying by Greenpeace and other NGOs, governments around the world signed the Stockholm Convention on Persistent Organic Pollutants in 2001. Similarly, Greenpeace's pressure was instrumental in the effective enforcement of Brazil's environmental crimes law since 1999 (Fuentes Véliz, 2007).

Initially relegated to observer status in international environmental law, NGOs have seen their role evolve to become active participants in drafting international treaties and instruments, particularly non-binding ones, thereby enriching their content. Their increasing involvement is essential not only for accelerating consensus on pressing environmental issues but also for strengthening the development of binding 'hard law' instruments, beyond the intermediate stages of 'soft law'. This evolution is particularly significant given the foundational principles of international environmental law: the precautionary principle, which calls for preventive measures even in the absence of scientific certainty; the polluter pays principle, which requires those responsible for environmental damage to bear the costs; and the principle of responsibility and cooperation, which obliges states to collaborate in preventing and remedying harm. However, despite these principles and the proliferation of treaties, challenges persist. Effective implementation is often undermined by insufficient resources and weak sanctions for non-compliance, while the integration of environmental concerns into broader economic and social development policies remains incomplete. Moreover, the law must constantly adapt to emerging threats, underscoring the need for NGOs to play a more central role in shaping and reinforcing international environmental governance.

In summary, the text acknowledges the growing importance of NGOs in the development of environmental law, shifting from limited participation to an active role in drafting legal instruments, both binding and non-binding.

Including non-governmental organisations in the negotiation process and the drafting of non-binding international instruments has been essential to the effectiveness of these declarations. Major NGO participation in the development of hard law instruments is essential because, given the current state of our planet, we must strive to reach a consensus on this issue directly, without the trial and error of soft law.

Recently, there has been a proliferation of environmental protection associations and organisations around the world, in recognition of the seriousness of the environmental crisis and the failure of various authorities and institutions to address it. These associations and organisations have found the judiciary to be a safe haven and stronghold, enabling them to compel authorities and private institutions to fulfil their environmental protection obligations. When considering climate cases, courts usually rely on national climate protection legislation, if it exists. Otherwise, they interpret and apply the provisions of international agreements.

3 The Urgenda Decision

In December 2019, the Dutch Supreme Court recognised for the first time that the Netherlands had a positive obligation to take measures to counter the real and immediate risk posed by global warming under Articles 2 (right to life) and 8 (right to private and family life) of the European Convention on Human Rights. The ruling established a causal link between greenhouse gas emissions from the Dutch state and global warming, as well as its current and future effects on Dutch territory. Based on this, the court upheld the injunction against the Dutch state to reduce its greenhouse gas emissions by at least 25% by the end of 2020 compared to 1990 levels (Gossement, 2019).

The Urgenda ruling played a pivotal role in establishing the basis for climate change litigation, in which the Dutch court ruled that the Dutch government must protect its citizens from the effects of climate change and reduce greenhouse gas emissions. This ruling came in response to a lawsuit filed by 900 Dutch citizens against their government for failing to adequately reduce emissions (Climate Change Litigation Databases, 2015).

Urgenda, an environmental organisation founded in 2008, filed an unprecedented lawsuit against the Dutch government, demanding stricter measures to reduce greenhouse gas emissions based on the state's commitments under the Paris Agreement. The organisation also filed a lawsuit against the government, demanding stricter measures based on the state's obligations under the European Convention on Human Rights, specifically Articles 2 and 8 relating to the right to life and the right to live with dignity. Urgenda explained that its goal in filing the lawsuit was to stimulate and accelerate the transition to a more sustainable society. The term 'sustainable society' is essentially human in nature. It recognises that human societies depend on the planet's natural resources and ecosystems for their existence and survival. It follows that these resources and ecosystems must be used and managed in a way that ensures the long-term survival of humanity and human societies and does not endanger them. A society in which economic activities lead to 'dangerous' climate change that threatens ecosystems and thus human societies around the world is, by definition, 'unsustainable'. Urgenda's claims, therefore, align with the pursuit of a 'sustainable society', which is its legal objective (Spijkers, 2018).

In 2015, the District Court of The Hague ruled that the government must reduce emissions by at least 25% by 2020 compared to 1990 levels. The court considered that failing to meet this commitment would pose a real threat to the fundamental rights of individuals, particularly those of future generations (De Schutter, 2020).

The government appealed against the ruling, arguing that the Netherlands' impact on global climate change was negligible and did not justify judicial intervention in public policy (De Schutter, 2020). However, both the Court of Appeal and the Supreme Court upheld the ruling, confirming that the size of a country's contribution to emissions does not exempt it from its responsibilities, as climate change is a collective threat requiring concrete national responses (Leijten, 2019).

The court concluded that global warming is caused by many factors and can only be solved through joint efforts. It would be totally unsatisfactory to allow the argument that the Netherlands' contribution to global greenhouse gas emissions is no more than 'a drop in the ocean' (i.e. no more than 0.35% (Leijten, 2019) of global emissions) to go unchallenged, since this applies to every country in the world, including the major powers. Ultimately, every contributor will be exempt, and no one will be held responsible for the dangerous rise in temperature of 2 °C or more that this could result in (Leijten, 2019).

The Urgenda case sets a unique legal precedent because the court granted citizens legal protection against a government that violated their right to protection from dangerous temperature rises (Minnerop, 2019). In doing so, the court placed itself in a position to review government policy on greenhouse gas emissions. It was clear that the court was aware of its role in developing the law. This ruling is merely a deterrent or risk-regulation mechanism (Minnerop, 2019).

4 The Emergence of Specialised Green Justice

International humanitarian law provides for the protection of the environment in two ways: firstly, under the general provisions, and secondly, under the additional special provisions. The general provisions relating to the impact of hostile operations on the natural environment are similar to those relating to civilian property, which may only be attacked if it is being used for military purposes. See Articles 35 and 55 of the First Additional Protocol to the four Geneva Conventions of 1949, which deal directly with environmental protection in armed conflicts, and Articles 54 and 56, which offer indirect protection.

In a case against the Czech Republic, the Prague court ordered the government to develop a more ambitious emissions reduction plan in line with its commitments under the Paris Agreement. In Scotland, Greenpeace sought a judicial review of the decision made by the UK Minister for Energy and Industrial Strategy and the North Sea Transition Authority to approve a new gas field in the North Sea in 2023, despite the climate damage that would be caused by burning the extracted gas. Such environmental lawsuits have become a key tool for compelling states to respect the environment, not merely emergency cases brought before the courts.

Traditionally, these lawsuits began as civil liability cases against fossil fuel projects, but they quickly evolved to focus more on environmental disputes in an attempt to pressure governments into developing effective environmental protection policies. This change in the judiciary's role, whereby it can now exercise control over governments to urge them to fulfil their obligations, has resulted in a steady increase in climate disputes. Approximately 2,341 climate disputes were recorded between June 2022 and 31 May 2023. Conversely, it has prompted a re-examination of judges' contribution to economic, social, and political governance.

The specific and complex nature of environmental disputes makes it difficult for ordinary judges to fulfil their role, and therefore the need for specialised green justice to deal with these disputes efficiently and effectively has become a reality of our time (Kaniaru, 2011).

Specialised courts primarily contribute to providing the multidisciplinary expertise suited to the complex nature of climate disputes. Secondly, they alleviate the burden on civil and administrative courts, thereby improving the efficiency with which these disputes are resolved (Kaniaru, 2011).

The emergence of international environmental justice is a relatively recent phenomenon, characterised by an increasing awareness of environmental inequalities and issues concerning global environmental protection (Kaniaru, 2011). This is manifested through social movements, political and legal initiatives, and a desire to hold those responsible, particularly states, accountable for environmental damage caused.

There are two types of environmental justice approach: local movements denouncing situations of environmental injustice, and global initiatives and public policies at the national or macro-regional level, which introduce social equity into environmental protection (Desponds, 2024) alongside the concept of sustainable development. The environmental justice movement emerged from the realisation that marginalised and low-income communities are disproportionately affected by pollution and environmental degradation (Desponds, 2024).

One of the main challenges is ensuring the effectiveness of international environmental law and guaranteeing that states comply with their obligations. The effectiveness of international environmental law can be enhanced by alliances with other bodies of law, such as human rights law. Judges have the power to ensure the effectiveness of international environmental law. They can do this in many ways, such as reminding political actors of their commitment to applying the law and specifying how it should be applied. These various trends form part of ongoing efforts to enhance the effectiveness of international environmental law, which requires constant updating. The law must play its part, as must the necessary economic and social changes. While the law can facilitate these developments, collective mobilisation across various fields of human activity is also required. In an urgent ecological context, the contributions gathered in this volume offer highly stimulating reflections on various pioneering forms of legal mobilisation in the field of environmental protection.

5 Conclusion

International environmental law is an essential tool for addressing global environmental challenges. However, its effectiveness depends on States honouring their commitments and adapting the law to changing realities. A comprehensive, integrated approach involving all stakeholders is necessary to ensure a sustainable future. Individuals and international organisations play a vital role in the application and effectiveness of international environmental law, and consequently, international law as a whole, as well as in establishing its binding nature. By participating in social movements that call for the establishment of international justice and the recognition of the right to a healthy environment, individuals can promote respect for international law, raise awareness of it, and sensitise communities to its importance, thereby putting pressure on states to comply with international law. International organisations contribute to the development, interpretation, application, and monitoring of international law, as well as the establishment of dispute settlement and enforcement mechanisms. Overall, the contributions of individuals and international organisations complement each other, enhancing the effectiveness of international law. While individuals help to create a social environment that promotes respect for the law, international organisations provide the necessary legal framework and mechanisms for its application and implementation.

References

Aguila, Y. (2015). *Renforcer l'efficacité du droit international de l'environnement—Devoirs des Etats, droits des individus* [Strengthening the effectiveness of international environmental law—Duties of States, rights of individuals]. Retrieved from https://think-tank.leclubdesjuristes.com/les-commissions/rapport-renforcer-lefficacite-du-droit-international-de-lenvironnement-devoirs-des-etats-droits-des-individus-2/#:~:text=Selon%20la%20Commission%20environnement%2C%20l,Des%20am%C3%A9liorations%20sont%20possibles%20

Climate Change Litigation Databases. (2015). *Urgenda Foundation v. State of the Netherlands.* Retrieved from https://climatecasechart.com/non-us-case/urgenda-foundation-v-kingdom-of-the-netherlands/

De Schutter, O. (2020). Changements climatiques et droits humains: l'affaire Urgenda [Climate change and human rights: The Urgenda case]. *Revue trimestrielle des droits de l'Homme, 123*(3), 567–608. Retrieved from https://droit.cairn.info/revue-trimestrielle-des-droits-de-l-homme-2020-3-page-567?lang=fr

Desponds, D. (2024). V. Les plateformes aéroportuaires: un accélérateur des injustices spatiales? [V. Airport platforms: an accelerator of spatial injustices?]. *Chemins du savoir,* 303–333. Retrieved from https://shs.cairn.info/article/HERM_DESPO_2024_01_0303

Fuentes Véliz, J. A. (2007). L'évolution du rôle des organisations non gouvernementales dans le droit de l'environnement [The evolving role of non-governmental organizations in environmental law]. *Revue Européenne de droit de l'Environnement, 11.4*, 401–30. Retrieved from https://www.persee.fr/doc/reden_1283-8446_2007_num_11_4_1980

Gossement, A. (2019, December 26). *Contentieux climatique] Affaire Urgenda c. État des Pays-Bas: la décision de la Cour suprême des Pays-Bas est-elle historique ou symbolique ? Droit de l'Environnement* [Climate Litigation] Urgenda v. State of the Netherlands: Is the Dutch Supreme Court's Decision Historic or Symbolic? Environmental Law]. Retrieved from https://www.gossement-avocats.com/blog/contentieux-climatique-affaire-urgenda-c-etat-des-pays-bas-la-decision-de-la-cour-supreme-des-pays-bas-est-elle-historique-ou-symbolique/#:~:text=Le%20sens%20de%20cette%20d%C3%A9cision%20est%20sans%20doute%20historique.,citoyens%20contre%20le%20changement%20climatique

Kaniaru, D. (2011). Launching a new environment court: Challenges and opportunities. *Pace Environmental Law Review, 29*, 626. https://doi.org/10.58948/0738-6206.1701

Leijten, I. (2019). Human rights v. Insufficient climate action: The Urgenda case. *Netherlands Quarterly of Human Rights, 37*(2), 112–118. https://doi.org/10.1177/0924051919844375

Tullio T. (2010). Les différends en droit international de l'environnement: règlement judiciaire et méthodes alternatives [Disputes in International Environmental Law: Judicial Settlement and Alternative Methods]. In *Le droit international face aux enjeux environnementaux, Colloque d'Aix-en-Provence* (Pedone, 2010) (pp. 433–450). Retrieved from https://pedone.info/sfdi/aix/SFDI_TREVES.pdf

Minnerop, P. (2019). Integrating the 'duty of care' under the European Convention on Human Rights and the science and law of climate change: the decision of The Hague Court of Appeal in the Urgenda case. *Journal of Energy & Natural Resources Law, 37*(2), 149–79. Retrieved from https://discovery.dundee.ac.uk/ws/portalfiles/portal/32465297/Author_Accepted_Manuscript.pdf

Peel, J., & Rebekkah, M. T. (2021). Recipe for success?: Lessons for strategic climate litigation from the Sharma, Neubauer, and Shell cases. *German Law Journal, 22*(8), 1484–1498. https://doi.org/10.1017/glj.2021.83

Shaw, M. (2025, July 29). *International law.* Encyclopaedia Britannica. Retrieved from https://www.britannica.com/topic/international-law

Setzer, J., & Catherine, H. (2025). *Global trends in climate change litigation.* snapshot. Retrieved from https://www.lse.ac.uk/granthaminstitute/publication/global-trends-in-climate-change-litigation-2025-snapshot/

Spijkers, O. (2018). The *Urgenda* case: A successful example of public interest litigation for the protection of the environment? In L. Kotzé (Ed.), *Courts and the environment* (pp. 305–344). Edward Elgar Publishing.

Dr. Rabiaa Bourhim holds a PhD in Public Law and Political Sciences. He is a member of the International Center for Strategic Research on Territorial Governance and the Research Center for Environment, Human Security & Governance (CERES). With expertise in government procurement, his research focuses on environmental change, conflict management, sustainability, and public policy. Dr. Bourhim has published numerous scientific articles, participated in conferences, and serves as a consultant to the parallel government of youth in Morocco.

Chapter 8
Defeating the Unenforceability of Ecocide

Orsolya Johanna Zsigmond-Sziebig

List of Abbreviations

AfCHPR:	African Court on Human and Peoples' Rights
CoE:	Council of Europe
ECD:	Environmental Crime Directive
ECHR:	European Convention of Human Rights
ECtHR:	European Court of Human Rights
EU:	European Union
HRBA:	Human Rights-Based Approach
IACtHR:	Inter-American Court of Human Rights
MS:	Member State
UN:	United Nations

1 Introduction

In recent years, humanity has been experiencing unprecedented changes in the environment, biodiversity, and living conditions. The drastic shifts in weather patterns, including droughts and floods, as well as the struggles faced by agriculture, all indicate that humanity must take immediate action to combat the adverse effects of climate change. Meanwhile, evolving armed conflicts once again jeopardise environmentalism, as wars have a detrimental impact on the environment and distract attention from the importance of environmental measures. Realising the paramount significance of environmental protection, a concept that is both new and yet "old"

O. J. Zsigmond-Sziebig (✉)
Department of International and European Law, Faculty of Law and Political Sciences, University of Szeged, Szeged, Hungary
e-mail: sziebig.orsolya.johanna@szte.hu

M. Mliless et al. (eds.), *Frontiers of Environmental Law Enforcement*,
https://doi.org/10.1007/978-3-032-10205-8_8

(Sziebig, 2020) has emerged in the past decade or so: ecocide. In such circumstances, it is crucial to reexamine legal concepts that can be used to hold those responsible for the most severe environmental damage accountable. The Russo-Ukrainian War, within the borders of continental Europe, has placed particular emphasis on ecocide, as the actual value of the ecological damage caused by the armed conflict, although incalculable, has been recorded and monitored from the outset (Ministry of Environmental Protection and Natural Resources of Ukraine, 2025). This chapter explores various possibilities regarding the future of ecocide legislation, with a focus on its enforceability. I want to provide an overview of the concept of ecocide, explain why its enforceability is questionable, and suggest ways to address this issue through creative legal thinking. The chapter is based on the fundamental hypothesis that, within the current regulatory framework, ecocide cannot be enforced directly at the international level in most cases. Instead, it serves as a justification and motivating factor for the adoption of specific new legislation. Another assumption is that in cases where ecocide is incorporated into particular national legislation, we can speak of actual legal enforcement under national jurisdiction. Between the international and domestic legislative approaches, it is essential to emphasise regional instruments, particularly those of the Council of Europe and the European Union. Among the sectors of international law, the international criminal law approach and the human rights aspects will feature in the discussion. The concept of ecocide and the modification or completion of the relevant legal instruments will be one of the core legal challenges in the coming decades, supported by several nations and organisations. In my opinion, ecocide is one of the most significant legal questions of this century, as it influences various other branches of law, including international and national environmental law, the law of armed conflicts, human rights law, and international criminal law.

Under the term "*environment*", the chapter includes both the natural and human-made aspects, reflecting the approach that the Stockholm Conference applied in 1972: "*Man is both creature and moulder of his environment, which gives him physical sustenance and affords him the opportunity for intellectual, moral, social and spiritual growth. In the long and tortuous evolution of the human race on this planet a stage has been reached when, through the rapid acceleration of science and technology, man has acquired the power to transform his environment in countless ways and on an unprecedented scale. Both aspects of man's environment, the natural and the man-made, are essential to his well-being and to the enjoyment of basic human rights the right to life itself.*" (United Nations, 1972, p. 3).

2 Methods

Although I have been researching ecocide for five years, new questions continue to arise. This is partly because ecocide increasingly appears in legislation, either as the subject of regulation or as a motivating factor. My research outcomes so far have concentrated on the history, definition, and legal embeddedness of ecocide; meanwhile, in this chapter, I aim to elaborate on the enforceability of ecocide. This chapter relies on the analysis of relevant articles and applies mainly qualitative research methods. The human rights and international criminal law aspects will be analysed using SWOT[1] analysis, highlighting the strengths and weaknesses of each approach. Due to the limited length of the work, the history and definition of ecocide are presented in a synthesis form, just for clarification and conceptualisation.

3 History of ecocide and its Definition

3.1 History of ecocide

The history of ecocide dates back to the USA's destructive war tactics applied during the Vietnam War, with particular regard to the intensive use of herbicides. The term "ecocide" was first documented in 1970, when botanist Professor Arthur W. Galston argued against environmentally destructive war tactics due to their impact on both ecological and human health (Zierler, 2011, p. 15). At the UN Conference on Human Environment, Swedish Prime Minister at the time, Olof Palme, also raised his voice against the outrageous Vietnam War Effort, labelling it as ecocide (Ecocide Law 2025a). Following up on the Conference, a working group (the Working Group on Crimes Against the Environment) was established to consider a draft ecocide convention that was also submitted to the United Nations in 1973. The convention has never been accepted to this day; however, in literature, it remains a relevant source for defining and explaining ecocide (Falk, 1973). The next chapter of ecocide's history strongly connects to the International Law Commission (ILC), the body responsible for codification within the UN. Between 1984 and 1996, the ILC addressed the issue of environmental crime as part of the draft Code of Crimes against the Peace and Security of Mankind. For a long time, Art. 26. was sentenced for the issue of environmental destruction and harm as the fifth crime against peace. Initially, the debate centred on the subjective element of the crime, specifically the assessment of acts that damage the environment, and, more precisely, whether such acts could be regarded as crimes of intent. As a result, in 1991, the wording was changed to "*wilful and severe damage to the environment*" (United Nations 1996a, pp. 15–29). In 1995, a special working committee was established, which published a report on its findings a year later. It identified three solutions for adapting environmental crime in the convention (United Nations 1996b).

Firstly, regulating environmental crime as a separate article in the future convention. Secondly, extending crimes against humanity to environmental crimes; or lastly, classifying the relevant actions as war crimes. In 1996, the article on environmental crime disappeared entirely from the draft, without the subject being put to a vote.

[1] SWOT: Strengths, Weaknesses, Opportunities, Threats.

From then on, the ILC could only decide under which crimes environmentally harmful acts would be classified. The result of this process was the adoption of the term "*widespread long-term and severe damage to the natural environment*", which later found its way to the Rome Statute (Human Rights Consortium, 2013). The modern history of ecocide started with the work of "Polly" Higgins, a Scottish international lawyer, who co-founded the STOP Ecocide Foundation with Jojo Mehta. Higgins aimed to amend the Rome Statute to include ecocide, and for this purpose, a definition was submitted to the United Nations. Higgins also published a book on ecocide, which includes various definitions and focuses on the history of environmental harm (Higgins, 2016). Some states have already legislated ecocide as part of domestic law, and due to the changes in EU law, the number of such countries is expected to grow in the upcoming years (Ecocide Law, 2025b).

3.2 Ecocide Definitions

The concept of ecocide has long been shrouded in uncertainty, as there has been no universally accepted scientific or legal definition. The highly expressive English term "ecocide" is a combination of two words: *"eco"* (from *"oikos"*, Greek), meaning ecosystem and *"cide"* (from *"caedere"*, Latin), meaning destruction, kill, or cut down (Robinson, 2021). Ecocide refers to large-scale environmental destruction in its most general sense. Due to the scope of the article, only the basic interpretations of ecocide are highlighted. The original submission of Polly Higgins contained the following definition: "*Ecocide is the extensive damage to, destruction of or loss of ecosystem(s) of a given territory, whether by human agency or by other causes, to such an extent that peaceful enjoyment by the inhabitants of that territory has been severely diminished*" (Higgins, 2016, pp. 62–63). The definition created by an Independent Expert Panel established by the STOP Ecocide Foundation can definitely be considered a milestone in the conceptualisation of the term. The expert panel defined ecocide as a recommendation towards the ICC to amend the Rome Statute with the following phrase: "*Unlawful or wanton acts committed with knowledge that there is a substantial likelihood of severe and either widespread or long-term damage to the environment being caused by those acts*" (STOP Ecocide Foundation, 2021, p. 4). In simpler terms, ecocide refers to severe environmental destruction that causes or is likely to cause extensive, long-lasting, or widespread damage to the environment and is punishable under both international and national law. Views among experts are divided on the subjective aspect of the offence, that is, whether only intentional acts or also negligent acts should be punishable. The environment should be interpreted as broadly as possible, including terrestrial, aquatic, marine, subsoil, and atmospheric environments.[2] It is also worth noting, as a final remark on the definition of ecocide, that there remains a debate among scholars regarding the subjective elements and the intent of the crime. Although the original definition

[2] On the definition of ecocide, see in detail: (Zsigmond-Sziebig, 2025)

by Falk contained a special intent,[3] the newer concepts have moved away from this understanding, as noted in the proposed article by the expert panel (Heller, 2021).

4 The Embeddedness of ecocide in International and Domestic Legal Spheres

The following Table 8.1 summarises the appearance of ecocide as a legal concept or subject of legislation, providing an overview of the different scopes examined. Additionally, these branches of law are highlighted in which ecocide can appear indirectly, serving as a rationale for legislation or a basis for decision. The different scopes are explained as part of the discussion.

4.1 International Law

As part of international law, the original "cradle" of ecocide was international criminal law, as it was already emphasised. Environmental matters indirectly adressed as part of war crimes in Article 8(2)(b)(iv): "*Intentionally launching an attack in the knowledge that such attack will cause incidental loss of life or injury to civilians or damage to civilian objects or widespread, long-term and severe damage to the natural environment which would be clearly excessive in relation to the concrete and direct overall military advantage anticipated*" (Rome Statute of the International Criminal Court, 1998 (2002)). The already quoted amendment to the Statute would establish a fifth crime against peace, integrating ecocide into the founding document of the ICC and providing the Court with jurisdiction to prosecute such crimes directly, based on the individual environmental crime. On the international level, a human rights-based approach might provide a creative way to remedy environmental harm. In these cases, HRBA is applicable, as environmental destruction often results in human rights violations as well. It is worth noting that, in the Inter-American sphere, the American Convention on Human Rights (1969/1978) does not explicitly mention environmental rights; however, the IACtHR has developed a strong environmental case law, with particular regard to the rights of Indigenous Peoples. A recent advisory opinion also strengthens the Court's willingness to establish a strong link between human rights protection and environmental matters (STOP Ecocide International, 2025). In the African sphere, the African Charter on Human and Peoples' Rights recognises the right to a satisfactory environment (Art. 24), and the AfCHPR has also developed an intensive environmental case-law.

---

[3] "*[ecocide] means any of the following acts committed with intent to disrupt or destroy, in whole or in part, a human ecosystem [...]*" (Falk, 1973, p. 21)

Table 8.1 Ecocide as a subject of legislation and its appearance in human rights courts[4]

Level of legislation		**Importance**	**Enforceability**	**Effect of legislation**
International sphere	*International criminal law and the International Criminal Court*	Indirectly addresses environmental crime as a form of war crimes, art. 8(2)(b)(iv) The proposed amendment (2025) would include ecocide as a fifth crime against peace.	Limited, but the Office of the Prosecutors is committed to emphasising the importance of environmental crimes.	There is no precedent, practically a "sleeping clause" of the Rome Statute.
			It would provide a solid legal basis for prosecuting environmental crimes.	The Rome Statute needs to be amended—a little reality.
	International human rights law	Applicability of human rights-based approach.	Through accepted human rights conventions and climate litigation.	It may provide a creative way to enforce ecocide in the human rights sphere.
Regional spheres	*Council of Europe*	No explicit recognition of environmental rights in the ECHR. New convention "Convention on the Protection of the Environment through Criminal Law" adopted in 2025—"particularly serious offences"	The ECtHR interprets the rights already declared by the ECHR and provides a remedy for environmental harm. The new legislation provides enforceable crimes, which are comparable to ecocide.	The Reykjavík process, which focused on and strengthened the work of the CoE, led to the acceptance of a new convention, calling for the recognition of ecocide. The signatories must ratify the convention and amend their domestic legislation.

(continued)

[4] Classification by Author, based on relevant literature and research.

Table 8.1 (continued)

Level of legislation		Importance	Enforceability	Effect of legislation
	European Union	The new environmental crime directive (EU) 2024/1203 specifically mentions ecocide in the preamble and introduces "qualified offences."	One of the goals is to improve the effectiveness of national enforcement chains. Enforceable crimes–actions or omissions that violate laws and can be addressed through legal proceedings by domestic authorities.	All the MSs must implement the Directive by May 2026. Emphasises the commitment of the European Union to sanction those who harm the environment.
	Inter-American sphere	No explicit recognition of ecocide, but strong environmental case-law at the IACtHR, including the rights of Indigenous Communities. Advisory Opinion No. 32 of the IACtHR (Climate Emergency and Human Rights, 2025).	Applicability of the human rights-based approach. Advisory opinions are not binding, but Advisory Opinion No. 32 is a crucial milestone that reflects the court's willingness to reflect environmentalism in its decisions.	Environmentally conscientious court practice—Pro-environment decisions. International obligation to prevent irreversible harm to the environment and the climate constitutes a jus cogens norm.
	African sphere	No explicit recognition of ecocide. The AfCHPR has environmental case-law as article 24 of the African Charter on Human and Peoples' Rights recognises the right to a satisfactory environment.	Applicability of the human rights-based approach.	Environmentally conscientious court practice—Pro-environment decisions.
State level (domestic legislation)		Several states have already penalised crimes which are comparable to ecocide.	Enforceable crimes.	Based on binding and enforceable legislation, perpetrators can be prosecuted and sanctioned.

4.2 Regional Appearance with Particular Regard to the European Sphere

The Council of Europe is a leading human rights organisation that was established by the European political integration, right after World War II. To this day, the CoE serves as a leading panel of treaty negotiations, resulting in the acceptance of several environmentally related treaties. Furthermore, the ECtHR has been focusing on environmental matters in its case law, filling the void in environmental rights under the European Convention of Human Rights (Sziebig, 2019, pp. 151–159). In recent years, ecocide has found its way into the discussion of the MSs of the CoE. Environmental protection and the correlation between a healthy environment and human rights are also emphasised in the Reykjavík process, as outlined in Appendix V, "*The Council of Europe and the environment*" (Council of Europe, 2023). In 2025, within the framework of the CoE, the MSs accepted a new "*Convention on the Protection of the Environment through Criminal Law*". The acceptance of the treaty is a significant milestone, as it is the first international, legally binding instrument to address a wide range of environmental crimes. The Council of Europe Parliamentary Assembly's Resolutions and Recommendations already reflected the importance of ecocide, and the new Convention explicitly mentions "*conducts comparable to ecocide*" that are "*particularly serious offences*". The latter term is based on Art. 31. as the following: "*destruction of, or widespread and substantial damage, which is either irreversible or long-lasting to an ecosystem of considerable size or environmental value or a habitat within a protected site or cause widespread and substantial damage which is either irreversible or long-lasting to the quality of air, soil, or water, such offences, leading to such devastating results, should constitute particularly serious criminal offences*" (Committee of Ministers, CoE, 2025, pp. 21–22). The new convention still needs to be ratified by the MSs, and it is hoped that it will gain more support than its forerunner. The CoE has already accepted a convention in a similar scope in 1998, but it attracted only 14 signatures, and only one state ratified the treaty. Despite the Convention's limited support, it has had a significant influence on EU legislation regarding environmental crimes (Bertram, 2025). Regarding the European Union, it is worth mentioning an EU citizens' initiative launched in 2012 called "*End Ecocide in Europe*". Its main objectives were to criminalise ecocide and to ensure that natural and legal persons are held accountable for committing ecocide (End Ecocide in Europe: A Citizens' Initiative to give the Earth Rights, 2012). In the European Union, the adoption of the new Environmental Crime Directive symbolises the growing attention on environmental criminal law and the commitment to ecocide (European Parliament and the Council, 2024). The new Directive mentions 'ecocide' in the preamble of the Directive. It also introduces the category of "*qualified criminal offences*" that are punished with more severe penalties than those applicable for other criminal offences defined in the legislation in Art. 3, para 3: "*Member States shall ensure that criminal offences relating to conduct listed in paragraph 2 constitute qualified criminal offences if such conduct causes: (a) the destruction of, or widespread and substantial damage which is either irreversible or long-lasting to, an ecosystem of considerable size or environmental value or a habitat within a protected site, or (b) widespread and substantial damage which is either irreversible or long-lasting to the quality of air, soil or water*".

It is clear that in the regulations of the CoE and the EU, as well as in the newly adopted convention and directive, ecocide appears only as a kind of 'motivating' factor, on which more severely punishable offences are based. A common feature of both regulations is that their provisions must be transposed into the national law of the Member States, meaning that the relevant laws of the participating countries must be aligned with the rules of the Convention, and all EU Member States must implement the Directive by spring 2026.

4.3 Domestic Embeddedness of Ecocide

A growing number of states are introducing specific ecocide legislation (Ecocide Law 2025a). Therefore, under the domestic law of such states, ecocide is a crime, and the perpetrators can be held responsible for environmentally destructive activities. The number of states with ecocide legislation is expected to increase, particularly in the European sphere, as they implement the new ECD and the Environmental Crime Convention of the CoE, which also catalyses the development of ecocide legislation. One of the main examples of ecocide legislation is the French regulation "Climate & Resilience Act", which was accepted in 2021. In April 2021, the French National Assembly passed a new law that would impose extremely harsh penalties on those who damage the environment. Offenders could face up to 10 years in prison (délit) and fines of up to €4.5 million (L'Assemblée nationale et le Sénat, 2021).

5 (Un)enforceability of Ecocide

In the upcoming subchapter, I aim to explore the intersection of international criminal law (Fig. 8.1) and international human rights law (Fig. 8.2) in the context of remedying environmental crimes. As it was mentioned earlier, environmental considerations can be included in different segments of law, including the human rights-based approach, climate litigation, and, more recently, the assessment of climate change as ecocide (Proedrou & Pournara, 2025). To provide a more comprehensible approach, the following simplified SWOT analysis outlines the benefits and drawbacks of each legal area. Each SWOT analysis focuses on the potential strengths, weaknesses, opportunities, and threats associated with the applied approach.

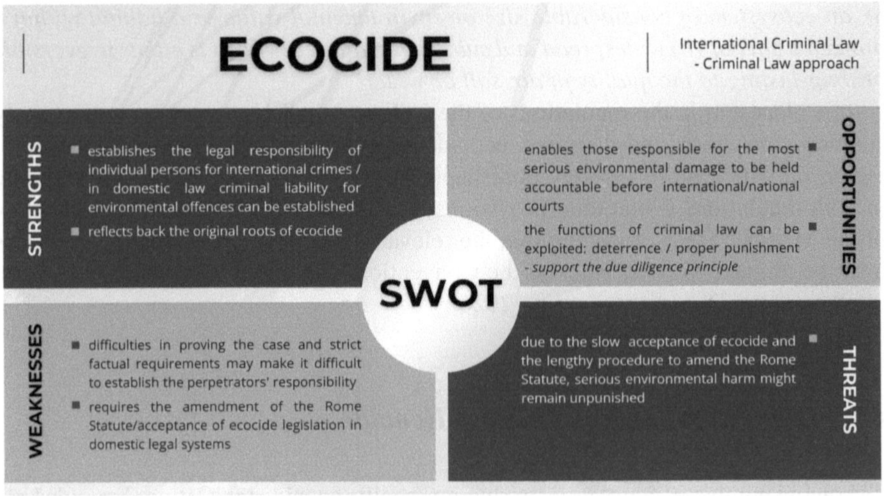

Fig. 8.1 SWOT Analysis—Criminal Law Approach

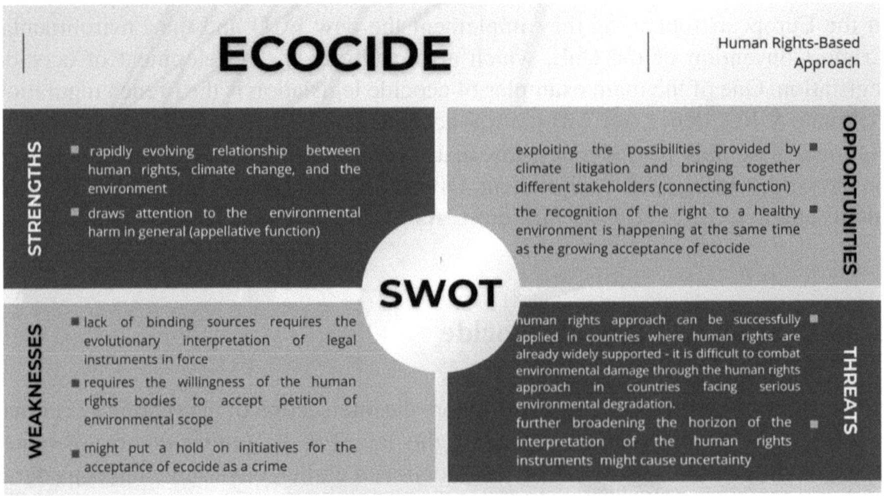

Fig. 8.2 SWOT Analysis—Human Rights-Based Approach

5.1 Obstacles and possibilities in enforcing ecocide

Enforcement under Criminal Law

Criminal law is often labelled as the last resort among applicable legal responses (Husak, 2004). Criminal law may only interfere with people's rights to the extent necessary, which is embodied at the level of criminal law principles by the requirements of *ultima ratio* (subsidiarity) and proportionality (Nagy, 2010, p. 46). Criminal law represents the most severe intervention in the individual sphere, and it can only be invoked when other legal instruments are insufficient. Of all the instruments of state control, it is indispensable for the enforcement of public interests and the maintenance of public order. Criminal law is only legitimate as a last resort for the most serious violations of the law (Nagy, 2010, p. 59). Therefore, establishing legal responsibility is possible only in the case of individuals who commit acts dangerous to society, and criminal penalties have to be imposed to remedy the actions. Criminal liability for environmental offences can also be established as ecocide if there is applicable legislation. The criminal law approach reflects the original roots of ecocide, which date back to the 1970s. The criminal law approach has a benefit, as it enables those responsible for the most serious environmental damage to be held accountable before international or domestic judicial bodies. Furthermore, the functions of criminal law can be exploited, including deterrence. It was highlighted in literature, that the acceptance of ecocide can serve as a deterrent, especially at the corporate level, as it adds a "personal risk" for company executives and may affect decision-making within the organisation (Hamilton, 2024).

One of the main issues of enforceability is a lack of binding ecocide legislation. As previously mentioned, prosecuting environmental crimes at the ICC is currently limited. Article 8(2)(b)(iv) of the Rome Statute is particularly challenging to prove, and its scope is limited, applying only to international armed conflict or non-international conflicts where there is a protracted armed conflict between the government and armed groups. Furthermore, the *actus reus* (physical element of the crime) must be "*widespread, severe and cause long-term environmental damage*". As part of the *actus reus*, the military advantage must also be examined; the damage must be disproportional to the overall military advantage anticipated. The act must be accompanied by a *mens rea*, a criminal intent (Gilman, 2011, pp. 454–455). As a result, the environmental element of war crimes established a challenging triple *mens rea* test. The test requires that the attack was launched intentionally, and the perpetrator knew that the anticipated environmental harm would be "*widespread, long-term, and severe*". Lastly, the anticipated damage was clearly excessive in relation to the concrete and the overall direct military advantage anticipated. According to some authors (Sharp, 1999, pp. 237–239), crimes against humanity would provide a more flexible framework for assessing environmental crimes. At the same time, crimes against humanity protect the civilian population, providing that the civilian population and individual civilians shall enjoy general protection against dangers arising from military operations. As a result, environmental crimes are connected indirectly to crimes against humanity. The possibility of prosecuting environmental crimes through crimes against humanity especially appeared in the context of the Indigenous communities of the Brazilian Amazon. This idea has been explored in literature (Grisafi, 2020) and practice (The Planet vs Bolsonaro (International Criminal Court October 12, 2021)), so far without any tangible results.

Overall, it is unlikely that the International Criminal Court will become the main forum for adjudicating environmental crimes in the future. This is despite the fact that the Policy Paper issued by the Office of the Prosecutors (OTP) promised more decisive action in relation to the adjudication of environmentally harmful acts. The real effect of the new OTP Policy Paper is questionable, but it has "*symbolic and precedential value; it suggests that environmental crimes are going to be taken seriously in the future*" (Patel, 2016, p. 197).

Another solution would be the amendment of the Rome Statute, as proposed originally by Higgins and the Independent Expert Panel established by the STOP Ecocide Foundation. An individual crime would serve as a legal basis to prosecute environmentally destructive acts at the ICC. Amending the Rome Statute is a lengthy procedure. As Professor Péter Kovács[5] also noted, it was already evident during the preparatory work for the ICC that states were divided on the jurisdiction of the International Criminal Court, especially regarding certain acts, such as the illegal trade of drugs or terrorism. The scope of international crimes extends beyond the jurisdiction of the International Criminal Court. For acts not covered by the Rome Statute, it is often the laws of individual states that provide the basis for proceedings, or hybrid courts may be established in practice. Given that even if the Assembly of States Parties adopts the amendment to the Rome Statute, all member states must ratify it; therefore, the expansion of the ICC's jurisdiction is less likely in the near future (Kovács, 2025, pp. 23–25). On 9 September 2024, Vanuatu, Fiji and Samoa proposed the recognition of "ecocide" as an international crime. The definition of ecocide corresponds to the International Expert Panel's terminology: "*unlawful or wanton acts committed with knowledge that there is a substantial likelihood of severe and either widespread or long-term damage to the environment being caused by those acts*" (Proposal—Independent Crime of Ecocide, 2024). The proposal also refers back to the new ECD of the European Union. One of the weaknesses of the international criminal law approach is the difficulty in proving cases, as strict factual requirements may make it challenging to establish the perpetrators' responsibility. Although based on the proposed definition of ecocide, human harm is not a requirement to hold the perpetrators responsible, the definition and its elements still generated heated debate among scholars.[6]

The most significant threat in relation to the applicability of the criminal law approach is the time aspect. Due to the slow acceptance of ecocide and the lengthy procedure to amend the Rome Statute, serious environmental harm might remain unpunished. As a result, alternative solutions must be explored to mitigate environmental harm.

Enforcement based on the Human Rights-Based Approach

Due to the shortcomings of the international criminal approach, it is worthwhile to explore the applicability of human rights instruments in environmental matters. As

[5] Péter KOVÁCS, judge at the International Criminal Court (mandate: 2015–2024), professor at the Péter Pázmány Catholic University of Budapest.

[6] On the debate, *see* (Zsigmond-Sziebig)

reflected earlier, the human rights bodies' case law presents a more environmentally cautious approach, especially in connection with environmental rights. On the general applicability of the HRBA, we can state that it can serve as a solution to remedy the lack of enforceable ecocide legislation and, in parallel, strengthen the protection of environmental rights. In my opinion, the connection between ecocide and human rights can be examined from two perspectives. Firstly, certain serious acts of environmental destruction that fall under the definition of ecocide also constitute serious violations of human rights. In a broader sense, this link calls for the development of a new human rights dimension in relation to climate change, based on intergenerational equality and the rights of future generations. The second aspect concerns human rights instruments, existing conventions, and judicial forums, as well as how these instruments can help remedy the damage caused by ecocide and whether it is possible to bring proceedings in this area before the competent forums. This question has led to climate litigation, which is gaining ground and attention on both the international stage and before national courts (Sziebig, 2024, pp. 80–82). In general, von Arnauld and Theilen identified five functions of human rights rhetoric: an appellative function, a contesting function, a connecting function, a triggering function and a jurisgenerative function (von Arnauld & Theilen, 2020). In my opinion, the benefits of the HRBA can be leveraged in relation to ecocide, including the possibilities provided by climate litigation. These procedures can bring together different stakeholders, such as states, civil society organisations, and the general public. The latter strongly resonates with the connecting function of HRBA.

On the other hand, there are some weaknesses in enforcing ecocide through human rights proceedings. Firstly, it requires the willingness of human rights bodies to accept petitions of an environmental scope, which sometimes still suffer from inadequacies, as exemplified by the Inuit petition (Petition To The Inter-American Commission on Human Rights Seeking Relief From Violations Resulting from Global Warming Caused By Acts and Omissions of the United States, 2005). Furthermore, the inclusion of ecocide rhetoric in the area of human rights might put a hold on initiatives for the acceptance of ecocide as a crime, as these procedures will reflect ecocide as a human rights issue and not a criminal law matter.

As threats, I have identified two angles. Firstly, the human rights approach can only be successfully applied in countries where human rights are already widely respected and upheld. Therefore, it is challenging to combat environmental damage through a human rights-based approach in countries facing severe environmental degradation, while also addressing the shortcomings of human rights protection. The ecocide debate resulted in a broader discussion about climate change and climate justice. It has been proven that lower-income countries, which are often developing states, are generally more vulnerable to the effects of climate change (Raus, 2025). Nevertheless, it is an excellent question whether ecocide can resolve the impunity gap for severe environmental damage and climate change, as to achieve this result, critical criminal law implementation is still needed. As a last resort, I believe we must acknowledge that further broadening the scope of interpretation of human rights instruments may cause uncertainty, as evolutionary interpretation is already a subject of debate (Vildigal, 2021).

6 Conclusion

As it is clear, the definition of ecocide and its constituent elements are themselves subjects for separate consideration, alongside questions of regulation and responsibility. The negative impact of severe environmental degradation on human rights is undisputed. The future of ecocide regulation is difficult to predict. At first glance, a significant part of the international community appears willing to commit to new regulations. This trend is likely to persist, as more countries experience the adverse effects of climate change. At the same time, the form of regulation remains unclear. I believe that the proposed amendment to the Rome Statute has paved the way for the future of ecocide. Under "pressure" from small island developing states, a proposal to amend the Rome Statute has now been initiated, but it will undoubtedly involve a lengthy ratification process. As it is clear from the analyses of different approaches, both criminal law and human rights-based approaches have the potential to remedy environmental harm. In my opinion, in the coming years we will see the parallel development of increasingly widespread enforcement of environmental rights and the expansion of legal regulation of ecocide. Given that these procedures are not mutually exclusive from a legal perspective, it is also possible that proceedings will be initiated in both types of forums following an event resulting in severe environmental damage. Of course, regarding individual procedures, it is essential to consider who may be a party to litigation in each forum, against whom proceedings can be initiated (states, companies, or individuals), and what the potential outcomes and legal implications might be. Although it is important to improve the eco-sensitiveness of the International Criminal Court (Killean, 2022), it is also crucial to explore "alternative" ways and procedures to remedy environmental harm and to put an end to ecologically destructive activities.

As Wendell Berry said, "*The Earth is what we all have in common*" (Berry, 1977, p. 128). The strengthening of the embeddedness of ecocide is clearly observable, indicating that the phenomenon has moved away from the original international criminal law approach and has found its way into other areas of international law. Today, ecocide primarily serves as a motivation for regulation, and there are few legal frameworks through which ecocide can be effectively enforced and used as a basis for establishing responsibility. Without cooperation, shared responsibility, and acceptance of the burdens involved in change, ecocide will remain merely an idea on the international community's agenda. Hopefully, states will find opportunities for cooperation so that ecocide can become a central element in establishing responsibility for environmental destruction in the coming decades.

References

Berry, W. (1977). The unsettling of America: Culture & Agriculture. Wendell Berry.

Bertram, D. (2025, June 4). *The first ecocide treaty*? Retrieved from EJIL: Talk! Blog of the. *European Journal of International Law.*. Retrieved from https://www.ejiltalk.org/the-first-ecocide-treaty/

Climate Emergency and Human Rights. (2025, July 3). Advisory Opinion 32 of 2025 (Inter-American Court of Human Rights).

Committee of Ministers, CoE. (2025, May 13–14). *Council of Europe Convention on the Protection of the Environment through Criminal Law—Explanatory Report.* Retrieved from https://search.coe.int/cm?i=0900001680b5cfab

Council of Europe. (2023, May 16–17). *Reykjavík Declaration: United around our values.* Retrieved from Reykjavík Summit of the Council of Europe: https://rm.coe.int/4th-summit-of-heads-of-state-and-government-of-the-council-of-europe/1680ab40c1

Ecocide Law. (2025a). *Ecocide/serious environmental crimes in national jurisdictions.* Retrieved from https://ecocidelaw.com/existing-ecocide-laws/

Ecocide Law. (2025b). *History.* Retrieved from https://ecocidelaw.com/history/

End Ecocide in Europe: A Citizens' Initiative to give the Earth Rights. (2012, August 2). Retrieved from https://www.europarl.europa.eu/meetdocs/2014_2019/documents/peti/dv/eciendecocideineurope_/eciendecocideineurope_en.pdf

European Parliament and the Council. (2024, April 11). *Protection of the environment through criminal law and replacing Directives 2008/99/EC and 2009/123/EC. OJ L (30.4.2024), 2024/1203.* https://eur-lex.europa.eu/eli/dir/2024/1203/oj/eng

Falk, R. E. (1973). Environmental warfare and ecocide-facts, appraisal, and proposals. *Bulletin of Peace Proposals, 4*(1), 80–96. https://doi.org/10.1177/096701067300400105

Gilman, R. (2011). Expanding environmental justice after war: The need for universal jurisdiction over environmental war crimes. *Colorado journal of international environmental law and policy, 22*(3), 448–471. Retrieved from https://scholar.law.colorado.edu/celj/vol22/iss3/5

Grisafi, L. (2020, November). Prosecuting international environmental crime committed against indigenous peoples in Brazil. *Columbia Human Rights Law Review Online, 5*, 26–59.

Hamilton, R. (2024, September 23). Why Criminalize Ecocide? Experts Weigh In. Retrieved from Just Security: https://www.justsecurity.org/100172/why-criminalize-ecocide-experts/

Heller, J. K. (2021, June 23). Skeptical thoughts on the proposed crime of "ecocide" (that isn't). Retrieved from Opinio Juris: https://opiniojuris.org/2021/06/23/skeptical-thoughts-on-the-proposed-crime-of-ecocide-that-isnt/

Higgins, P. H. (2016). *Eradicating Ecocide.Laws and Governance to Stop the destruction of the planet* (2nd ed.). Shepheard-Walwyn.

Human Rights Consortium. (2013). The ecocide project 'ecocide is the missing 5th crime against peace. (A. Gauger, M. P. Rabatel-Fernel, L. Kulbicki, D. Short, & P. Higgins, Eds.) Retrieved from https://sas-space.sas.ac.uk/4830/1/Ecocide_research_report_19_July_13.pdf

Husak, D. (2004). The criminal law as last resort. *Oxford Journal of Legal Studies, 24*(2 SUMMER), 207–235. https://doi.org/10.1093/ojls/24.2.207

Killean, R. (2022). *Reparation in the Aftermath of Ecocide., (pp. 1–12). An International Crime of Ecocide: New Perspectives, Symposium.* doi: https://doi.org/10.2139/ssrn.4315496.

Kovács, P. (2025). *The jurisprudential practice of the international criminal court—As I saw it.* Central European Academic Publishing. https://doi.org/10.54171/qms.2025.kpjpicc

L'Assemblée nationale et le Sénat. (2021). Law No 2021-1104 on the fight against climate change and the reinforcement of resilience in the face of its effects (LOI n° 2021-1104 du 22 août 2021 portant lutte contre le dérèglement climatique et renforcement de la résilience face à ses effets). Retrieved from Légifrance: https://www.legifrance.gouv.fr/jorf/id/JORFTEXT000043956924

Ministry of Environmental Protection and Natural Resources of Ukraine. (2025). *EcoZagroza.* Retrieved from https://ecozagroza.gov.ua/en

Nagy, F. (2010). *General part of Hungarian criminal law* [A magyar büntetőjog általános része] (2nd ed.). HVG-Orac.

Patel, P. (2016). Expanding past genocide, crimes against humanity, and war. *Loyola University Chicago Law Review, 14*(2), 175–197.

Petition To the Inter-American Commission on Human Rights Seeking Relief from Violations Resulting from Global Warming Caused by Acts and Omissions of the United States, Petition No. P-1413-05 (Inter-American Commission on Human Rights 2005). Retrieved from https://climatecasechart.com/non-us-case/petition-to-the-inter-american-commission-on-human-rights-seeking-relief-from-violations-resulting-from-global-warming-caused-by-acts-and-omissions-of-the-united-states/

Proedrou, F., & Pournara, M. (2025). Exploring representations of climate change as ecocide: Implications for climate policy. *Climate Policy, 25*(2), 269–282. https://doi.org/10.1080/14693062.2024.2368859

Proposal—Independent Crime of Ecocide. (2024). Proposal submitted by Vanuatu to amend the Rome Statute. ICC, Assembly of States Parties, ICC-ASP/23/26, 1 December 2024. Retrieved from https://asp.icc-cpi.int/sites/default/files/asp_docs/ICC-ASP-23-26-ENG.pdf

Raus, E. (2025, January 16). Ecocide: A new avenue for climate justice? In *Retrieved from world's youth for climate justice*. Retrieved from https://www.wy4cj.org/legal-blog/ecocide-a-new-avenue-for-climate-justice

Robinson, D. (2021, July 16). *Your Guide to Ecocide: Part* 1. Retrieved from Opinio Juris: https://opiniojuris.org/2021/07/16/your-guide-to-ecocide-part-1/

Rome Statute of the International Criminal Court. (1998 (2002)). *UNTS* Vol. 2187 (p. 3).

Sharp, P. (1999). Prospects for environmental liability in the international criminal court. *Virginia Environmental Law Journal, 18*(2), 217–243.

STOP Ecocide Foundation. (2021, June). *Independent Expert Panel for the Legal Definition of Ecocide: Commentary and Core text. Retrieved from STOP Ecocide Foundation*: https://static1.squarespace.com/static/5ca2608ab914493c64ef1f6d/t/67f539d588e9544792700921/1744124385734/SE%2BFoundation%2BCommentary%2Band%2Bcore%2Btext%2B2025.pdf

STOP Ecocide International. (2025, July 3). *Inter-American court of human rights: Preventing destruction of nature is a global legal duty.* Retrieved from STOP Ecocide International: https://www.stopecocide.earth/bn-2025/inter-american-court-of-human-rights-preventing-nature-destruction-a-global-legal-duty

Sziebig, O. J. (2019). European international organizations (Council of Europe, OSCE) and European regional cooperation (V4, Nordic council). In A. Szalai & Z. Fejes (Eds.), *Interstate relations* (pp. 151–171). Iurisperitus. Retrieved from https://acta.bibl.u-szeged.hu/69100/1/opera_iurisprudentiae_002_002_151-171.pdf

Sziebig, O. J. (2020, November 6–7). Ecocide, the new yet old international crime—shifting approaches and possible outcomes. The international biennial scientific conference organised by the West University of Timişoara.

Sziebig, O. J. (2024). The crime of ecocide through human rights approach: The "universal" right to a healthy environment as a driving force calling for ecocide legislation. *ACTA HUMANA: Hungarian Centre for Human Rights Publications, 12*(2), 75–89. https://doi.org/10.32566/ah.2024.2.5

The Planet vs Bolsonaro (International Criminal Court October 12, 2021). Retrieved from https://climatecasechart.com/wp-content/uploads/non-us-case-documents/2021/20211012_14633_na.pdf

United Nations. (1972, June). Declaration of the United Nations Conference on the Human Environment. Retrieved from https://docs.un.org/en/A/CONF.48/14/Rev.1

United Nations. (1996a, March 27). *Document on crimes against the environment, prepared by Mr. Christian Tomuschat, member of the Commission.* Retrieved from Draft code of crimes against the peace and security of mankind (Part II)—including the draft statute for an international criminal court: https://legal.un.org/ilc/documentation/english/ilc_xlviii_dc_crd3.pdf

United Nations. (1996b). Yearbook of the International Law Commission. Retrieved from https://legal.un.org/ilc/publications/yearbooks/english/ilc_1996_v2_p1.pdf

Vildigal, G. (2021). Evolutionary interpretation and international law. *Journal of International Economic Law, 24*(1), 203–219. https://doi.org/10.1093/jiel/jgaa035

von Arnauld, A., & Theilen, J. T. (2020). A topical perspective on the functions of claiming a 'human right to …'. In A. von Arnauld, K. von der Decken, & M. Susi (Eds.), *The Cambridge handbook of new human rights: Recognition, novelty, rhetoric* (pp. 34–50). Cambridge University Press. https://doi.org/10.1017/9781108676106.004

Zierler, D. (2011). *The invention of ecocide: Agent Orange, Vietnam, and the scientists who changed the way we think about the environment*. University of Georgia Press.

Zsigmond-Sziebig, O. J. (2025). *Ecocide—an evolving legislative concept in the light of care for creation*. L'Harmattan and the Pázmány Péter Catholic University. https://doi.org/10.56037/978-963-646-394-6

Orsolya Johanna Zsigmond-Sziebig is a senior lecturer at the Faculty of Law and Political Sciences at the University of Szeged, in the Department of International and European Law. She obtained her PhD in 2019 and an LLM in 2017. Since 2014, she has been teaching various subjects related to International and EU Law. Her research focuses on International and European Environmental Law, particularly wildlife crime, ecocide, and human rights. To date, she has published more than twenty research articles, four books, and fourteen book chapters, and has participated in numerous conferences both in Hungary and abroad. Her book "Ecocide - An Evolving Legislative Concept in the Light of Care for Creation" was published in 2025. She was honoured with a certificate of recognition from the Rector in 2022 and was named the best lecturer by students in the same year.

Chapter 9
Prevention of Environmental Crime in the Context of Promoting Public Policies and Environmental Education

Mihaela-Corina Bucur

1 Introduction

Humanity is simultaneously facing two major crises—the climate crisis and the environmental crisis—both of unprecedented scale and urgency. Climate change is a global challenge that requires a responsible approach and concrete action at the global, regional, national, and local levels. The scientific community has highlighted the devastating effects that climate change and environmental degradation have on human life and well-being. The main challenge of the 21st century is to transition to a development model that protects the environment and balances economic and social considerations with efforts to combat climate change and ensure the sustainable use of natural resources. This challenge requires an urgent change of direction. Systemic solutions and multi-level action are necessary: international cooperation, firm commitments from states, and the involvement of all state institutions, as well as the private sector, academia, and citizens. Changes in lifestyle, from diet to energy use, as well as changes in production, consumption, and transport methods have become an immediate necessity. Ecocide (ecological crime) is a largely hidden crime; very few acts are discovered, reported, investigated administratively or criminally, or punished by authorities, and the sanctions remain insignificant. This is despite the fact that environmental crimes rank fourth among international illegal activities, outranked only by drug trafficking, counterfeiting of currency and art, and human trafficking, and show a steady annual growth of 5-7%. They have become some of the most profitable forms of organized crime worldwide, significantly affecting both the environment and human health. Education that promotes a sustainable lifestyle is essential focusing on developing eco-social skills and familiarizing people with the natural and socio-economic problems caused by climate change, as well as ways to improve responses to it.

M.-C. Bucur (✉)
Law, Victimology, Romanian Society of Victimology, Bucharest, Romania
e-mail: prof.bucur@gmail.com

M. Mliless et al. (eds.), *Frontiers of Environmental Law Enforcement*,
https://doi.org/10.1007/978-3-032-10205-8_9

2 Dimensions of the Phenomenon

According to a Report by the United Nations Environment Programme (UNEP) and Interpol (June 2016)[1], environmental crime has reached a record high, with estimated losses of $258 billion, including: forest crimes and illegal logging: $51–152 billion; fishing crimes: estimated at $1124 billion, corresponding to 11–26 million tons diverted from legal trade and contributing to overfishing; illegal mining: $12–24 billion.

In the European Union (EU), annual revenues from the illicit trafficking of non-hazardous waste reach €1.3–10.3 billion, from hazardous waste trafficking €1.5–1.8 billion, and the volume of illegal trade in rare flora and fauna species threatened with extinction is €20 billion annually, constituting the second largest illegal market in the world, after drugs.

Between 2018–2021[2], police operations in the EU led to the arrest of 829 people, with total seizures amounting to €8.3 million, consisting of: 167,452 tons of waste, with the illegal waste market generating €4–15 billion annually; 1,174 tons of e-waste; one ton of smartphones contains one hundred times more gold than one ton of gold ore; 32,007 m^3 of timber, with illegal logging causing deforestation whose economic value is €6 billion annually, representing 15-30% of global timber trade; 639 ivory objects; elephant poaching is increasing the species' risk of extinction, with ivory obtained from Africa and trafficked through Europe to Asia; 4.6 tons of European eel juveniles – 350 million juveniles annually, worth €3 billion, with specimens sold mainly in Asia for consumption; 355 reptiles, valued at €800,000; thousands of reptile species worldwide are threatened by trafficking, primarily used in the fashion industry or as pets.

Ecological crime/Environmental crime/Ecocide can be conceptualized as a socially dangerous act, committed deliberately, consisting of: polluting the natural or artificial environment; hindering the prevention, reduction, or elimination of pollution; endangering the life and health of people, animals, and plants;[3] or causing significant economic damage that jeopardizes environmental quality and constitutes a threat to society's interests in environmental protection. The term "ecomafia" refers to the perpetrators of these crimes when they operate using methods typical of organized crime.

3 Concept of Environmental Policies

Environmental policy represents the totality of objectives and priorities, regulatory methods, and implementation instruments designed to ensure the sustainable use of natural resources and to prevent, reduce, or mitigate environmental degradation.

---

[1] https://eur-lex.europa.eu/legal-content/RO/TXT/PDF/?uri=CELEX:52021PC0851& from= EN

[2] https://www.consilium.europa.eu/ro/infographics/eu-fight-environmental-crime-2018-2021/

[3] Duțu, M. Introducere în Dreptul Penal al Mediului. Editura Hamangiu, București, 2013, p. 87.

The basis of the European Union's (EU) environmental policy is the Environmental Action Programs (EAPs). Their legal basis is found in Article 192, paragraph 3, of the Treaty on the Functioning of the EU. To date, eight EAPs have been developed, with the eighth, adopted on March 29, 2022, forming the basis of environmental policies until 2030[4].

Environmental policy instruments are basic mechanisms available to democratic governments and supranational structures that influence the attitudes and behavior of the population and economic actors, helping achieve established objectives and goals. Three types of instruments have been developed: legislative, technical, and economic-financial, as well as other support instruments responding to new trends and strategies for environmental protection[5].

European regulations are shaped by the European Green Deal[6], aiming to achieve climate neutrality in all relevant European policies and strategies, across all sectors of activity, as well as by the New EU Strategy on Adaptation to Climate Change (2021).[7] Its main objectives are: improving knowledge about the impact of climate change, intensifying adaptation planning and climate risk assessment, accelerating adaptation measures, increasing the resilience of the European continent to climate change.

4 Areas of Action of the European Union

Based on the assessment of the threat posed by serious crime and organized crime in the EU–2021, presented by Europol, the members of the Council of Europe (CE) have identified 10 crime priorities for the period 2022–2025: high-risk criminal networks; cyberattacks; human trafficking; child sexual exploitation; migrant smuggling; drug trafficking; fraud, economic and financial crimes; organized property crime; firearms trafficking; and environmental crimes. This was one of the priorities of the Successfully Weaponizing Illicit Wildlife Cases in Europe (SWIPE) project, in which the World Wildlife Fund (WWF) collaborates with several non-governmental organizations (NGOs) and partners from 11 European countries to increase the number of illegal activities detected and punished, as well as to strengthen cross-border cooperation, law enforcement, and prosecution.

The CE document discusses serious and organized crime activities that will be combated through the European Multidisciplinary Platform Against Criminal Threats (EMPACT). This is a multidisciplinary EU initiative based on information and concrete data, which aims to address the main criminal threats facing the EU. It is led

[4] https://www.consilium.europa.eu/ro/press/press-releases/2022/03/29/council-adopts-8th-environmental-action-programme/

[5] Petrescu-Mag, R-M. and Petrescu-Mag, I.-V. Legislation and Environmental Policies. Bioflux Publishing House, Cluj-Napoca, 2022.

[6] https://ec.europa.eu/info/strategy/priorities-2019-2024/european-green-deal_ro

[7] https://ec.europa.eu/clima/eu-action/european-green-deal/european-climate-pact_ro

by the Member States and involves many bodies working closely together, including law enforcement agencies, EU institutions and agencies, public and private organizations, as well as relevant countries and organizations outside the EU. EMPACT includes both preventive and repressive measures, as well as operational and strategic actions.

The purpose of cooperation between EU states regarding environmental crimes is to combat criminal networks involved in all their forms, especially those with the ability to infiltrate legal commercial structures or to establish their own businesses to facilitate their crimes.

The main objective of WWF and all partners who have joined forces in this initiative is to contribute to the reduction of crimes against European wildlife by filling gaps in wildlife crime databases; providing relevant technical knowledge to prosecutors and control authorities; and raising awareness of the negative effects of these crimes on security, public health, and the environment.

The EU's areas of action focus on the following issues: air, chemicals, the circular economy, industry, the marine and coastal environment, nature and biodiversity, noise, plastics, soils and land, sustainable development, the urban environment, waste and recycling, and water.

Air[8]—Air quality has deteriorated considerably due to human activities causing polluting emissions, particularly from industry, energy production, residential heating, agriculture, and transport. Air pollution is the main environmental health problem in the EU, causing serious diseases such as asthma, cardiovascular problems, and lung cancer, with vulnerable groups being the most affected. Excessive air pollution with nitrogen compounds and acid rain also harms the environment and ecosystems, being costly for the economy as it leads to lost working days and high healthcare costs. In the EU, 300,000 premature deaths occur annually due to fine particles, with the EU target being to reduce these deaths by at least 55% by 2030; in addition, the annual economic cost of air pollution is at least €330 billion. The EU's clean air policy aims to improve ambient air quality and combat air pollution to protect the environment and human health. To address air pollution and achieve the EU's zero pollution vision for 2050, the EU has a comprehensive clean air policy based on three pillars: ambient air quality standards, the reduction of air pollution emissions, and emission standards for key pollution sources. More precisely, EU policies aim to reduce the number of premature deaths and illnesses caused by air pollution and reduce pressure on ecosystems and biodiversity.

Circular economy[9]—Biodiversity loss is 90% caused by resource extraction and processing; up to 80% of the environmental impact of products is determined in the design phase, and the current rate of circular material use in the EU is only 11.8%. The EU aims to transition to a circular economy to make Europe cleaner and more competitive, a transition that will reduce pressure on natural resources, create sustainable growth and jobs, and is essential to achieve the EU's climate neutrality

[8] https://environment.ec.europa.eu/topics/air_en

[9] https://environment.ec.europa.eu/topics/circular-economy_en

target for 2050 and to halt biodiversity loss. Moving away from the linear "take-make-use-dispose" model and transitioning to a regenerative growth model is essential to keep resource consumption within planetary boundaries. In a circular economy, the value of products, materials, and resources is maintained in the economy for as long as possible, and waste generation is kept to a minimum. The current linear economy continuously increases its demands on limited natural resources. By using and consuming in a more circular way, the environmental impact of human economic activities, including their effects on biodiversity, can be substantially reduced. To accelerate the EU's transition, the CE adopted the New Circular Economy Action Plan in March 2020. The circular economy helps decouple economic growth from resource use. It protects Europe's natural resources, stimulates sustainable growth, and enables the EU to reduce its consumption footprint while aiming to double its circular material use rate in the next decade. The circular economy will enable a healthier planet by reducing pollution and pressure on natural resources such as water and land use. It will reduce emissions, helping the EU become the first climate-neutral continent, create new business opportunities and quality local jobs, and enable more resilient value chains.

Nature and biodiversity[10]—Humanity depends on nature for food, air, water, energy, and raw materials. Nature and biodiversity make life possible, offer health and social benefits, drive our economy, and are our best ally in tackling the climate crisis. Currently, nature is being destroyed worldwide faster than ever before. Ecosystems and the services they provide are under severe pressure from urban expansion, intensive agriculture, industrial pollution, invasive species, and climate change. In the EU, 81% of habitats are in poor condition, and 18% of land is part of the "Natura 2000" protected area - the largest coordinated network of protected areas in the world, established by the EU, with the EU target being that by 2030, 30% of land and sea will be protected. As part of the European Green Deal, the EU Biodiversity Strategy for 2030 is a comprehensive, ambitious, and long-term plan to put biodiversity on the path to recovery by 2030, protecting nature and reversing ecosystem degradation, with specific commitments and actions based on existing nature laws. The Birds and Habitats Directives ("Nature Directives") constitute the EU's oldest environmental laws, forming the backbone of EU biodiversity policy and representing the legal basis for "Natura 2000."

Environmental Noise (Noise Pollution)[11] is a major environmental health problem in Europe, caused by various sources, including road, rail, and airport traffic, industrial activity, and construction. It can be substantially reduced by implementing effective and cost-efficient measures. According to the World Health Organization, environmental noise is the second largest environmental cause of health problems after air pollution. Prolonged exposure to noise can lead to serious diseases, including cardiovascular conditions, reduced cognitive performance in children, severe stress-related disorders, sleep disorders, and tinnitus. In the EU, 12,000 premature deaths occur annually due to long-term exposure to ambient noise, 22 million people suffer

[10] https://environment.ec.europa.eu/topics/nature-and-biodiversity_en

[11] https://environment.ec.europa.eu/topics/noise_en

from chronic stress-related disorders, and 12,500 students suffer from learning disabilities caused by noise. The effects of noise exposure are also costly for the economy, as they lead to fewer productive working days, strain healthcare systems, and reduce property values. EU rules address noise pollution from primary sources, external sources, and underwater noise, which affects marine animals. The EU's noise policy has harmonized procedures to assess the effects of noise on human health. It aims to monitor environmental noise across Europe and contribute to achieving the zero pollution vision by 2050. The CE has also proposed a policy objective aiming to reduce the proportion of people disturbed by transport noise by 30% by 2030.

Plastics[12]—are very important in our economy and daily lives, but they can have serious negative effects on the environment and human health. Almost 26 million tons of plastic waste are generated in Europe each year, approximately 80% of marine litter is plastic, and 87% of Europeans are concerned about the environmental impact of plastic products. The EU's policy on combating plastic pollution aims to protect the environment and human health by reducing marine litter, greenhouse gas emissions, our dependence on imported fossil fuels; transforming the way plastic products are designed, produced, used, and recycled in the EU, accelerating the transition to a sustainable, circular, and resource-efficient plastic economy, supporting safer and more sustainable consumption and production models for plastics, creating new opportunities for innovation, competitiveness, and jobs, stimulating change, and setting a global example. Specific rules and objectives apply to certain areas, including single-use plastics, plastic packaging, microplastics, and soon, biodegradable and compostable plastics.

Soils and Land[13]—represent the foundation of our health and wealth, forming the basis of the economy and prosperity, supporting our cultural heritage and landscapes. Life on Earth itself depends on healthy soils, a finite and non-renewable natural resource that stores, filters, and transforms many substances, including water, nutrients, and carbon, and is essential for climate change mitigation and adaptation, for agricultural production and food security, nature and biodiversity conservation, and vital for achieving the key objectives of the European Green Deal. In the EU, 95% of our food comes from soil, but land and soils are degrading dramatically in Europe and worldwide, with 60-70% of soils being unhealthy, and €50 billion lost each year due to soil degradation. To protect human health and the environment, EU policy aims to improve soil health for the benefit of people, food, nature, and climate by: taking measures and actions to protect, restore, and sustainably use soils; presenting a vision to achieve healthy soils by 2050; improving the soil monitoring framework in Europe; developing the knowledge base and supporting soil research; and raising awareness of the vital importance of soils. The CE will present a new law on soil health, which will provide a comprehensive legal framework, giving it the same level of protection as exists for water, the marine environment, and air in the EU.

---

[12] https://environment.ec.europa.eu/topics/soil-and-land_en

[13] https://environment.ec.europa.eu/topics/soil-and-land_en

Urban Environment[14]—Europe is becoming increasingly urbanized: over 70% of Europeans live and work in cities or suburbs. The quality of life in urban areas greatly depends on environmental quality, air and water quality, access to nature and biodiversity, noise pollution levels, and, increasingly, the ability to moderate temperatures low as global warming progresses. Cities consume the most resources and face many environmental challenges. However, they also offer opportunities to tackle environmental problems. People live and work closer together, manage and share resources efficiently, apply circular economy principles, reduce reliance on private motorized transport, and live in more energy-efficient buildings. In the EU, one in five people lives in areas with harmful noise levels, and 300,000 premature deaths occur annually due to fine particles in the environment. The EU's urban policy works to ensure strong protection of the environment, nature and biodiversity, human health, and a good quality of life in urban areas. The CE funds a wide range of projects and initiatives focusing on environmental laws regarding air, water, the conservation and restoration of green spaces, improving waste management, strengthening the circular economy, and ensuring that citizens are not exposed to excessive noise. Cities play a crucial role in achieving the objectives of the European Green Deal and delivering a low-carbon, resource-efficient, sustainable, and resilient society. EU cities and their local governments are therefore essential partners in implementing environmental laws, policies, and programmes.

Waste and Recycling[15]—An average of five tons of waste is produced per person in Europe each year; only 38% of waste in the EU is recycled, and over 60% of household waste ends up in landfills in some EU countries. Environmentally sound waste management and the use of secondary materials are key elements of EU environmental policy. The European Green Deal supports the transition to a circular, modern, resource-efficient, and competitive economy by extracting high-quality resources from waste whenever feasible. To support this transition, several EU waste laws are set to be reviewed. Certain categories of waste require specific approaches, so the EU has numerous laws addressing different types of waste. The Waste Framework Directive provides the legal framework for waste treatment and management in the EU and introduces a "waste hierarchy" that prioritizes waste management methods. The EU's waste policy aims to protect the environment and human health and sets objectives to improve waste management, stimulate innovation in recycling, and limit waste disposal.

Clean Water[16] is the driving force of life, an essential resource for people and nature and for climate regulation, and is crucial for the economy, agriculture, and energy production. Water faces many problems, including pollution from industrial chemicals, pesticides, nutrients, and pharmaceuticals, as well as climate change. In the EU, only 40% of surface waters achieve good ecological status, while almost 85% of bathing water sites are of very good quality, and over 90% of urban wastewater is treated in accordance with EU standards. The EU's water policy is one of the

---

[14] https://environment.ec.europa.eu/topics/urban-environment_en

[15] https://environment.ec.europa.eu/topics/waste-and-recycling_en

[16] https://environment.ec.europa.eu/topics/water_en

cornerstones of environmental protection in Europe. The Water Framework Directive provides the main framework and objectives for water policy in Europe. The main goal of EU rules is to ensure that all Europeans have access to sufficient, good quality water, that water is managed sustainably in the long term, that water pollution is reduced, and that freshwater and saltwater ecosystems are protected.[17]

5 European Union Policies

The Zero Pollution Action Plan for Air, Water, and Soil (2021) is based on three pillars: ambient air quality standards, reduction of air pollution emissions, and emission standards for key pollution sources. Its aim is to reduce the number of premature deaths and illnesses caused by air pollution and to lessen the pressure of pollution on ecosystems and biodiversity. The Sustainable and Smart Mobility Strategy envisages reducing emissions in the transport sector by 90% by 2050 through the widespread adoption of sustainable transport solutions.

- The **European Climate Law (2021)** has legislated climate neutrality objectives and the intermediate target of reducing greenhouse gas emissions, including the proposal to reduce greenhouse gas emissions by 55% by 2030 compared to 1990 levels, taking into account the transformation of the energy system, including buildings, transport, and industry, through decarbonization.
- The **EU's noise pollution policy** aims to monitor environmental noise in Europe to contribute to achieving the zero-pollution vision for 2050. EU rules address noise pollution from primary sources, external noise sources, and underwater noise, which affects marine animals, with the EU's policy objective being to reduce the proportion of people disturbed by transport noise by 30% by 2030.
- The **Water Framework Directive** provides for all Europeans to have access to sufficient, good quality and sufficient water, long-term sustainable management, reduction of water pollution, and protection of freshwater and saltwater ecosystems.
- The **EU Soil and Land Strategy (2021)** includes measures and actions for the protection, restoration, and sustainable use of soils, achieving healthy soil by 2050, improving the soil monitoring framework, developing the knowledge base and supporting soil research, and increasing awareness of the vital importance of soils. The CE will present a new law on soil health, which will provide a comprehensive legal framework for its protection.
- The **EU Biodiversity Strategy for 2030**[18] aims to restore biodiversity by 2030, including specific commitments and actions to protect nature and reverse ecosystem degradation, based on existing nature laws. "Natura 2000" is the largest coordinated network of protected areas in the world, built by the EU, with its target of protecting 30% of land and sea by 2030.

---

[17] https://eur-lex.europa.eu/legal-content/.

[18] https://ec.europa.eu/environment/strategy/biodiversity-strategy-2030_ro

- The **New EU Forest Strategy for 2030 (2021)** emphasizes supporting the socioeconomic functions of forests; protecting, restoring, and expanding them; ensuring sustainable management, and strategic monitoring for adaptation to climate change and increasing forest resilience.
- The **EU's circular economy policy**. The New European Bauhaus Initiative (2021)[19] translates the European Green Deal into a cultural and creative dimension. This initiative aims to accelerate the transformation of different economic sectors and provide everyone with access to low-carbon, circular goods. The circular economy helps decouple economic growth from resource use, protects Europe's natural resources, and stimulates sustainable development. It will reduce the consumption footprint, double the circular material use rate over the next decade, and decrease pollution, emissions, and pressure on natural resources, helping the EU become the first climate-neutral continent.
- The **EU's plastics policy** aims to reduce marine litter, lower greenhouse gas emissions and decrease dependence on imported fossil fuels. It also seeks to transform how plastic materials are designed, produced, used, and recycled in the EU. The policy supports the transition to a sustainable, circular, and resource-efficient plastic economy, encourages safer and more sustainable consumption and production models, creates new opportunities for innovation and competitiveness, and stimulates change across the sector.
- The **EU's urban environment policy** funds multiple projects and initiatives focused on environmental laws, and cities play a crucial role in achieving the objectives of the European Green Deal by providing a low-carbon, resource-efficient, sustainable, and resilient society.
- The **EU's waste and recycling policy** aims to improve waste management, stimulate innovation in recycling, and limit waste disposal. The European Green Deal favors the transition to a circular, modern, resource-efficient, and competitive economy by extracting high-quality resources from waste. The Waste Framework Directive constitutes the EU's legal framework for waste treatment and management and introduces an order of preference for management, called the "waste hierarchy."
- The **European Research Strategy for 2020–2024 (2020)**[20] has placed research and innovation at the heart of European ambitions to achieve full climate neutrality and resilience by 2050.

[19] https://ec.europa.eu/commission/presscorner/detail/en/ip_21_4626

[20] https://ec.europa.eu/info/sites/default/files/rtd_sp_2020_2024_en.pdf

6 Prevention Of Ecological Crime and Environmental Protection

The prevention and combating of ecological crime, along with environmental protection, have become permanent priorities for public authorities, specialists, and the general public alike. These efforts aim to safeguard human living conditions and public health, support economic and social interests, and ensure the sustainable development of society.

Prevention activities in environmental protection are an integral part of environmental policy. The social value of environmental crime prevention involves not only reducing the scope of criminal law but also minimizing the costs of environmental crimes and mitigating their harmful effects.

The general task of environmental protection[21] is to safeguard environmental factors, create a legislative framework for identifying and eliminating the pollution causes, define measures for prevention, mitigate its effects, and implement these measures in line with the public interest. All this can only be achieved depending on the economic and financial capacity of society, with international cooperation, and taking into account real priorities, which concern human health and life[22].

International cooperation in the field of environmental protection requires the establishment of ecological monitoring compatible with advanced global standards. Monitoring is necessary to determine existing environmental conditions, assess health risks before and after certain situations; the efficiency of pollution control equipment; compliance with agreements and regulations; punishment for violating current regulations; and the responsibilities of partners, legal entities, and individuals.

The development of a new criminological prevention model involves the study of the determinants of environmental crime. The main drivers of this type of criminality are economic. Capitalism, with its focus on capital accumulation and maximizing profits while minimizing costs, creates conditions for environmental pollution. When environmental costs are absent, economic actors can increase profitability at the expense of the environment, while the state benefits financially in the short term by exploiting natural resources unsustainably.

The current challenge is revising the system for implementing new technologies in production and management. These activities are expensive even at the state level and hinder the implementation of the necessary reform, as well as the creation of a new integrated model for pollution prevention.

The main way to force entities not to pollute the environment is to make pollution unprofitable. For this, it's necessary to promote the introduction of the latest technologies, environmental insurance, increase sanctions for environmental violations, and raise the value of environmental monitoring indicators.

---

[21] Stoleru, E. (2020). Ecological Criminality: Criminological Dimensions and Prevention Measures. In: Legea şi Viaţa, pp. 86-89.

[22] Lupan E., Environmental Law. General Part. Elementary Treatise. Vol. I. Bucharest: Lumina Lex Publishing House, 1996, pp. 7-8.

The direct consequences of environmental degradation include natural disasters, increased population morbidity, and reduced life expectancy. The main problem in combating environmental pollution is that the primary polluters are industries and the economic activity of large enterprises. By comparison, the contribution of individuals to pollution is minimal.

Environmental protection requires setting a general objective for environmental quality and the active involvement of enterprises to find the most efficient ways to address tasks related to combating harmful environmental factors and protecting human ecosystems.

Today, a number of new technologies are available that can lead to a reduction in atmospheric air pollution: the implementation of renewable energy sources and non-polluting technologies. The basic principle is the protection of the environment, communities, nations, and of humanity as a whole.

Protecting the balance of human ecosystems includes, first of all, planning local areas to reduce health risks, developing public and private transport, creating green spaces and adequate conditions for rest and work. Solutions must be future-oriented, incorporating aspects related to risk prevention, such as anticipating climate change or progressively reducing dependence on fossil fuels. Local initiatives to solve certain problems can create new issues in other areas and may contradict policies at the national or regional level. Clearly defining objectives and targets, assuming responsibilities, implementing progress monitoring procedures, public consultation, verification of results, auditing and reporting are crucial for the effective implementation of environmental protection measures.

Ultimately, ensuring environmental safety should be carried out jointly by environmental agencies and law enforcement authorities. In the process of implementing measures to prevent ecological crimes, it is important to remember that these crimes represent a real threat to the biological foundations of human life, health, and species development.

7 Environmental Education and Climate Change

An important mission in environmental protection is educating new generations about nature protection. Environmental education is a methodology where people gain information about the environment and acquire learning, skills, values, experiences, and passion, all of which will empower them to act–separately and collectively–to address current and future environmental issues[23].

Climate change and environmental degradation are major challenges that will affect future generations. Addressing them must involve radical changes at both economic and social levels. Education is one of the pillars for enhancing the response to climate change, by changing human behavior, with a view to protecting nature

[23] Stoleru, op. cit., pp. 86-89.

and resources. It is recognized that climate education plays a fundamental role in society's behavioral and psychological adaptation to climate change.

Environmental education and climate change education are essential instruments for social change, promoting public awareness and concrete actions to create a sustainable future. Thus, the mission of the 21st century is to achieve sustainable development, using education as a strategy to ensure sustainability.

Children and young people must be educated about climate change and the environment. This education should promote a sustainable lifestyle and develop eco-social competencies. It should help them understand natural and socioeconomic problems caused by climate change and equip them to respond effectively. The goal is to raise awareness of climate change and environmental problems, with children being able to act as messengers to their families and communities, as well as being directly involved in efforts to stop environmental degradation. At the same time, education is essential for developing public policies and implementing measures for environmental protection and combating climate change.

Surveys among adults indicate the need to raise awareness of the seriousness of climate change and to increase the degree of personal responsibility in combating it. Thus, according to a Eurobarometer measuring European citizens' attitudes towards climate change, published in July 2021, an average of 49% of Europeans believe that climate change is the most serious problem facing the world[24]. At the same time, most Europeans believe that "for every trade agreement concluded by the EU, there should be a guarantee that the highest standards regarding climate, environmental, and labor protection are met"[25]. There is, therefore, a need to increase the level of understanding of the climate change phenomenon through both formal education and the involvement of media and social networks.

The educational system should prepare well-informed and responsible citizens. They should be capable of contributing to climate change mitigation, environmental protection, transitioning to a sustainable economic growth model, and adopting a sustainable lifestyle. Educational institutions should become strategic partners in building a sustainable future, and students and teachers should benefit from all necessary support to understand the seriousness of the climate and environmental crisis, become aware of their causes and effects, but also of ways to improve society's response to these challenges.

It is imperative to cultivate respect for the environment, as part of the values promoted and developed through education, and for the formal education system to train citizens who are responsible towards society and the environment. Research and innovation are necessary to support climate actions, food security, at-risk communities, adaptation to climate change, and effective responses to emergencies generated by climate change.

---

[24] European Commission (2021). Climate Change - Eurobarometer survey. Available at: https://europa.eu/eurobarometer/ surveys/detail/2273

[25] European Commission (2021). Standard Eurobarometer 94 - Winter 2020-2021. Available at: https://europa.eu/ eurobarometer/surveys/detail/2355

Environmental and climate change education should be integrated throughout the educational system. Additionally, schools and educational facilities must meet European and global standards for sustainable development.

Education on climate change and the environment addresses European priorities by increasing students' understanding of the factors that cause environmental degradation and climate change. It also encourages their active participation in practical environmental protection activities. Therefore, the objective of environmental and climate change education must be the development of competencies[26] (knowledge, skills, and attitudes) necessary for adapting to climate change and protecting the environment.

The three essential components for operationalizing environmental education are: infrastructure, curriculum, and pedagogy, aiming for: elements of sustainable school infrastructure (superior energy efficiency through thermal insulation, renewable energy generation systems, green space planning, selective waste collection and disposal, etc.) and an environmentally and climate-responsible curriculum that incorporates content on and for the environment.

Students develop the capacity to reflect and act to protect the environment, develop basic competencies for a circular economy based on the sustainable use of natural resources, and strive to adopt a lifestyle and a culture of sustainability that promotes the protection of biodiversity and the restoration of natural ecosystems.

Climate change and environmental education aim to contribute to shaping individual, group, and systemic behaviors that prevent to the degradation of nature and the environment, and to empower people to contribute to problem-solving and systemic changes. In addition, the purpose of climate change and environmental education is to contribute to achieving long-term sustainability goals, to support relevant actors in the fight against the serious effects of pollution, environmental degradation, and climate change, and to increase their capacity for effective intervention. This education thus aims at orienting towards change and action, taking into account the following dimensions: (1) Culture of complexity; (2) Capacity for action; (3) Co-responsibility for the environment.

The values, behaviors, and attitudes targeted by environmental and climate change education are: empathy, a connection with nature; care and compassion for all forms of life; respect, interest, and appreciation for nature and the services it offers to people; responsibility regarding the use of natural resources; responsibility in managing one's own decisions that have environmental and climate impact, including in daily consumption behaviors; involvement in solving environmental and climate problems; participation and civic responsibility.

The objective of climate change and environmental education must be directed towards the development of competencies through which students contribute to: (1) relevant actions to combat climate change and reduce its impact; (2) actions to adapt to climate change and ensure resilience to climate change; (3) protecting, restoring, and promoting the sustainable use of terrestrial ecosystems, rivers, lakes, seas, and

[26] ISE Description of European Key Competencies. Available at: https://www.ise.ro/wpcontent/uploads/2015/04/ Competente-cheie-europene.pdf

marine resources, and halting biodiversity loss; (4) ensuring sustainable consumption and production models.

Students should develop the following capacities: (1) understand the basic concepts about the environment and climate change, including causes, effects, and mitigation measures; (2) develop systemic thinking at a global level and understand how climate change connects with energy, irresponsible exploitation of natural resources, pollution, the food industry, the extractive industry, the energy industry, social justice, food waste, natural disasters, and other sectors; (3) explore/investigate and interact with their environment; (4) make daily decisions considering their impact on the planet and adopt environmentally responsible behaviors; (5) adapt to extreme weather events and respond to natural disasters; (6) understand environmental legislation, relevant actors, and their roles in order to take personal actions; (7) participate in developing public policies and new technologies to combat climate change; (8) understand climate change, environmental care, and sustainable development in the context of global events; (9) recognize the importance of collective action and engage in projects, debates, and workshops to advance sustainable development goals.

The main goal of education and professional training is to develop competencies in children, young people, and adults. This includes fostering sensitivity to human and environmental issues, moral and civic values, and respect for nature and the natural, social, and cultural environment, starting from the premise that the education system must play a fundamental role in preparing active citizens who take responsibility for the environment and combat climate change.

Sustainable development should strengthen society's ability to adapt and its resilience capacity to combat dangers related to climate change and natural disasters. This requires integrating mitigation and adaptation measures into strategies, policies, and planning, as well as improving education and awareness about climate change. In the context of these changes, the following objectives are priorities: improving education, awareness, and human and institutional capacity regarding climate change mitigation, adaptation, impact reduction, and early warning; intensifying efforts to achieve the transition to a "green," low-carbon economy resilient to climate change, and integrating climate change adaptation measures into vulnerable economic, social, and environmental sectors, in accordance with EU policies.

The new European legislative framework, developed around the main objective of the European Green Deal, to achieve climate neutrality by 2050 and to fulfill the Sustainable Development Goals of Agenda 2030, requires a rethinking of the public policy framework to align actions and measures included in normative acts.

Implementing a strategy and action plan means that both legal and strategic frameworks must evolve over time. These updates should align with the proposed objectives and consider the European legislative context for climate change and environmental protection.

8 Legislative Updates in Romania and Europe

In 2025, environmental crimes are defined as illegal acts that cause damage to the environment. These include the illegal collection and disposal of waste, pollution of water, air, or soil, and trafficking of protected species. In Romania, Government Emergency Ordinance No. 195/2005 regulates these offenses and provides corresponding sanctions. For example, Article 98 penalizes both the illegal transport of hazardous substances and the intentional pollution of water. In the European Union, a recent directive[27] expands the list of environmental offenses and introduces harsher penalties. These include illegal timber trafficking, the unlawful recycling of polluting ship components, and serious violations of chemical legislation.

Law No. 56/2025 introduces significant changes to waste management legislation, explicitly including air among the environmental components whose deterioration can trigger criminal liability. Government Emergency Ordinance No. 92/2021 governs cases in which violations of waste regulations constitute criminal offenses. Authorities aim to: expand the scope of criminal sanctions for acts affecting air quality and ensure clearer alignment with EU standards, which promote an integrated approach to environmental protection. This addition aligns with fundamental principles of environmental law, including the precautionary principle and the "polluter pays" principle, as well as with European Directives on environmental crimes.

Damage to air quality is often a direct consequence of illegal practices, such as the incineration of waste in unauthorized spaces, illegal disposal of volatile or toxic substances, or non-compliant transport. Previously, acts that produced harmful emissions into the atmosphere were not treated with the same degree of criminal severity due to the lack of explicit reference to air. Law no. 56/2025 addresses this legislative gap. It provides clearer mechanisms for environmental authorities, law enforcement, and prosecutors to investigate and penalize offenses affecting the atmosphere.

9 Conclusion

Environmental crimes constitute a serious and growing problem that manifests in multiple forms, including from air, water, and soil pollution, to crimes that accelerate climate change, the destruction of fish stocks, deforestation, the depletion of essential natural resources, and the disappearance of commercially valuable wild species. These crimes can also have detrimental impacts on the economies and security of several nations and, in extreme cases, could even jeopardize the existence of a country or its people. Furthermore,organized crime networks have been shown to play a significant role in pollution and wildlife crimes. Recently, these networks have expanded their activities to include the trafficking of plant and animal species,

[27] Directive of the European Parliament and of the Council on the protection of the environment through criminal law and replacing Directives 2008/99/EC and 2009/123/EC

in addition to the export of hazardous waste and radioactive materials. This highlights the close link between organized crime and ecological crime. Environmental crimes have often been considered non-specific business crimes. They require special business knowledge only when they cause or risk causing substantial financial losses typically committed by business people in the exercise of their professions or functions. Given the scale of ecocide and the numerous unintentional ecological problems, it is necessary to adopt and apply robust legislative provisions and sanctions at both national and international levels, ensuring their consistent implementation. These measures should be coherently harmonized and coordinated. Additional measures should include combating tax havens and increasing financial support proportionate to the gravity of the threats posed by ecological crime; promoting sustainable development, economic and fiscal responsibility, and providing alternative livelihoods for those involved at the end of the criminal chain. It is also essential to increase education and awareness among children and young people regarding sustainable development and responsible environmental behavior. The state of the environment depends on each of us. We must recognize that negligence toward the environment can have serious consequences and that we have the power to protect and preserve life on Earth.

10 General Conclusion

10.1 Toward a Reimagined Future of Environmental Law Enforcement

This volume has taken the reader through a multidimensional journey across the frontiers of environmental law enforcement. Beginning with the tools of forensic science and artificial intelligence that allow investigators to detect and reconstruct environmental crimes, the chapters then examined the global networks of transnational organized crime that exploit ecological resources with impunity. From there, the focus narrowed to the national level with the case of Morocco's Environmental Police, highlighting how institutional innovation can both expand enforcement capacity and reveal the limits of fragile governance structures.

Education emerged as another critical layer. By rethinking pedagogy and embedding ethics, civic responsibility, and ecological justice into curricula, enforcement is framed not only as a matter of legal authority but also as a cultural practice that shapes future generations. At the same time, chapters on armed conflict reminded us that law remains weakest where it is most needed: in times of war, when ecosystems are weaponized, poisoned, or destroyed alongside human communities.

Legal imagination and reform also featured prominently. Analyses of green international law underscored both the potential and the fragmentation of existing frameworks, while the debate on ecocide illustrated how new legal categories can expand accountability and redefine the boundaries of international criminal law. The final chapters drew attention to prevention and policy, emphasizing that enforcement cannot rely solely on punitive mechanisms but must be complemented by education, community engagement, and cultural shifts toward sustainability.

Taken together, these contributions affirm that environmental crime is no longer peripheral—it is central to the ecological, legal, and geopolitical realities of the twenty-first century. They also reveal that enforcement, in its narrow sense, is insufficient. What is required is a dynamic approach that integrates science and law, global and local institutions, punitive action and preventive strategies, justice for people and justice for ecosystems.

As editors and contributors, we do not claim to provide definitive answers. Rather, this volume offers a framework for rethinking enforcement as a living practice—one that adapts to changing contexts, confronts systemic inequalities, and remains anchored in justice. If environmental law is to retain meaning, it must be enforced. And if it is to retain integrity, that enforcement must be fair, innovative, and forward-looking. The future of environmental protection will depend on our capacity to combine these principles—in laboratories, in courts, in classrooms, and in international fora alike. This volume is offered as one step toward that horizon.

Mohamed Mliless
Meknes, Morocco
25/08/2025

References

Directive of the European Parliament and of the Council on the protection of the environment through criminal law and replacing Directives. (2008). 2008/99/EC and 2009/123/EC

Duțu, M. (2013). *Introduction to Environmental Criminal Law*. Hamangiu Publishing House.

European Commission. (2021a). Climate Change-Eurobarometer survey. https://europa.eu/eurobarometer/surveys/detail/2273

European Commission. (2021b). Standard Eurobarometer 94-Winter 2020–2021. https://europa.eu/eurobarometer/surveys/detail/2355

Government Emergency Ordinance. No. 195/2005 on environmental protection

ISE Description of European Key Competencies. (2022). https://www.ise.ro/wpcontent/uploads/2015/04/Competente-cheie-europene.pdf

Law No. 56/2025 for the amendment of Article 66, paragraph (1^2), letter c) of Government Emergency Ordinance No. 92/2021 regarding the waste regime

Lupan, E. (1996). *Environmental Law. General Part. Elementary Treatise*. Vol. I. Lumina Lex Publishing House.

Petrescu-Mag, R.-M., & Petrescu-Mag, I.-V. (2022). *Legislation and Environmental Policies*. Bioflux Publishing House.

Stoleru, E. (2020). Ecological Criminality: Criminological Dimensions and Prevention Measures. In *Legea și Viața* (pp. 86–89).

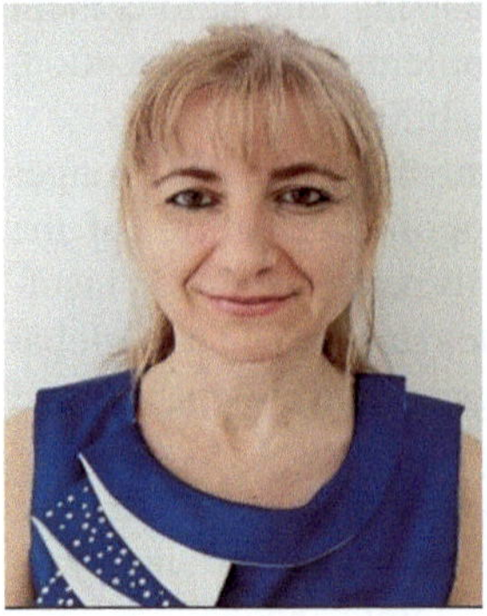

Mihaela-Corina Bucur holds a Master's degree in Educational Management and has academic qualifications in Theology, Music, Medicine, and Law. She is a tenured teacher in health-care education, manager of the Asanka Sante Medical Clinic, and also practices medicine at Bucharest Airports, Romania. In addition, she is a certified mediator, jurist, and founding member of the Romanian Society of Victimology. Bucur collaborates with the Romanian Academy of Scientists and has published research across multiple fields, including medicine, pedagogy, theology, and legal studies. Her scholarly work addresses a wide range of topics such as ecocide, ecological education, cellular therapies, public policies, medical malpractice, bullying, cyber-bullying, mobbing, domestic violence, doping, and restorative justice.

General Conclusion

Toward a Reimagined Future of Environmental Law Enforcement

This volume has taken the reader through a multidimensional journey across the frontiers of environmental law enforcement. Beginning with the tools of forensic science and artificial intelligence that allow investigators to detect and reconstruct environmental crimes, the chapters then examined the global networks of transnational organized crime that exploit ecological resources with impunity. From there, the focus narrowed to the national level with the case of Morocco's Environmental Police, highlighting how institutional innovation can both expand enforcement capacity and reveal the limits of fragile governance structures.

Education emerged as another critical layer. By rethinking pedagogy and embedding ethics, civic responsibility, and ecological justice into curricula, enforcement is framed not only as a matter of legal authority but also as a cultural practice that shapes future generations. At the same time, chapters on armed conflict reminded us that law remains weakest where it is most needed: in times of war, when ecosystems are weaponized, poisoned, or destroyed alongside human communities.

Legal imagination and reform also featured prominently. Analyses of green international law underscored both the potential and the fragmentation of existing frameworks, while the debate on ecocide illustrated how new legal categories can expand accountability and redefine the boundaries of international criminal law. The final chapters drew attention to prevention and policy, emphasizing that enforcement cannot rely solely on punitive mechanisms but must be complemented by education, community engagement, and cultural shifts toward sustainability.

Taken together, these contributions affirm that environmental crime is no longer peripheral—it is central to the ecological, legal, and geopolitical realities of the twenty-first century. They also reveal that enforcement, in its narrow sense, is insufficient. What is required is a dynamic approach that integrates science and law, global and local institutions, punitive action and preventive strategies, justice for people and justice for ecosystems.

M. Mliless et al. (eds.), *Frontiers of Environmental Law Enforcement*,
https://doi.org/10.1007/978-3-032-10205-8

As editors and contributors, we do not claim to provide definitive answers. Rather, this volume offers a framework for rethinking enforcement as a living practice—one that adapts to changing contexts, confronts systemic inequalities, and remains anchored in justice. If environmental law is to retain meaning, it must be enforced. And if it is to retain integrity, that enforcement must be fair, innovative, and forward-looking. The future of environmental protection will depend on our capacity to combine these principles—in laboratories, in courts, in classrooms, and in international fora alike. This volume is offered as one step toward that horizon.

Mohamed Mliless
Meknes, Morocco
25/08/2025

Index

M. Mliless et al. (eds.), *Frontiers of Environmental Law Enforcement*,
https://doi.org/10.1007/978-3-032-10205-8

MIX
Papier aus verantwortungsvollen Quellen
Paper from responsible sources
FSC® C105338

If you have any concerns about our products,
you can contact us on
ProductSafety@springernature.com

In case Publisher is established outside the EU,
the EU authorized representative is:
Springer Nature Customer Service Center GmbH
Europaplatz 3, 69115 Heidelberg, Germany

Printed by Libri Plureos GmbH
in Hamburg, Germany